DANGER
UXB

Also by James Owen

A Serpent in Eden

DANGER
UXB

THE HEROIC STORY OF THE
WWII BOMB DISPOSAL TEAMS

JAMES OWEN

Little, Brown

LITTLE, BROWN

First published in Great Britain in 2010 by Little, Brown

Copyright © James Owen, 2010

The moral right of the author has been asserted.

A CIP catalogue record for this book
is available from the British Library.

Hardback ISBN 978-1-4087-0195-9
C-format ISBN 978-1-4087-0255-0

Typeset in Bembo by M Rules
Printed and bound in Great Britain by
Clays Ltd, St Ives plc

Papers used by Little, Brown are natural, renewable and
recyclable products sourced from well-managed forests and certified
in accordance with the rules of the Forest Stewardship Council.

Mixed Sources
Product group from well-managed
forests and other controlled sources
www.fsc.org Cert no. SGS-COC-004081
© 1996 Forest Stewardship Council

FSC

Little, Brown
An imprint of
Little, Brown Book Group
100 Victoria Embankment
London EC4Y 0DY

An Hachette UK Company
www.hachette.co.uk

www.littlebrown.co.uk

For 'Felix' – those of every generation
engaged in Bomb Disposal

ACKNOWLEDGEMENTS

All works of history are mosaics. Many people have given generously of their time, expertise and advice over the last couple of years, so enabling me piece by piece to assemble this book.

I am particularly grateful to those who allowed me to interview them about their experiences in Bomb Disposal, or who shared memories of their fathers' lives and work. I should like to thank for these opportunities Colonel Stuart Archer, Deirdre Strowger, Professor Dick Hudson, the Earl of Suffolk, Maurice Howard, Ricky Hards, June Daughtrey, John Emlyn Jones, Oswin Kent and Elga La Pine.

Captain Sandy Sanderson kindly allowed me to consult the archives of the Explosive Ordnance Disposal Technical Information Centre, and patiently answered my novice questions about fuzes. WO1 Colin Rae instructed me in the chemistry of high explosive, and Tina McKenna helpfully arranged my visit to 33 Engineer Regiment (EOD). Steve Venus shared his vast knowledge of wartime BD, and Chris Ransted that of its casualties. Jo Wisdom at St Paul's Cathedral put me in touch with Peter Boalch, who freely gave me the fruits of his own long research into the UXB there.

Robin Bennett and Gary Woodman-Simmons provided much-needed information about BD veterans, and Didy Grahame and Terry Hissey points of contact for winners of gallantry awards.

Marion Hebblethwaite started me on some valuable lines of enquiry, and I commend her biographical dictionary of recipients of the George Cross to all those stirred by tales of courage. Paul Hughes produced detail about Con Stevens, and Hans Houterman repeatedly conjured up natal dates and Army numbers that I was unable to find elsewhere.

Ken Abraham delved into the Newry & Mourne Museum for more about Max Blaney. I was also much aided and guided by the collections staff of the Imperial War Museum, under the watchful stewardship of Roderick Suddaby, and by those of the National Archives, the National Army Museum, the British Library, the Royal Institution, the Royal Society, the Royal Engineers Museum, Library and Archive, the Second World War Experience Centre, the Historical Disclosures section of the Army Personnel Centre, EODTIC and the London Library. I must also thank Michelle Miles for her swift and efficient research in the Bundesarchiv, Freiburg, and for her translation of material from it. Jamie Crowe, in Sydney, and Rosemary Falls, in Dubai, provided refuges in which to write when a volcano was doing its best to prevent me from delivering the final pages of the book.

I am tremendously grateful for the support and forbearance of my publisher Richard Beswick and my copy-editor Zoe Gullen, and for the perspicacity and encouragement of my literary agent Julian Alexander. Despite their best efforts, all judgements expressed in the book are my responsibility alone, as are any errors.

CONTENTS

LIST OF ILLUSTRATIONS

Foreword

A NATION ON GUARD

On the evening of 23 September 1940 the people of Britain gathered around their radios to listen to a broadcast by their king. Speaking in the deliberate manner that had become familiar to them, each phrase a struggle with his stammer, he praised their 'courage and cheerfulness' in the face of the 'Blitz' unleashed by the Germans two weeks earlier, above all against London. This was a new kind of war, one that for the first time placed the British population on the front line and sought through bombing to break their morale. 'Tonight,' the king told them, 'we are a nation on guard.'

Then he made an unexpected announcement. Now that the once-distant carnage of the battlefield had come to the streets and factory yards, he had decided to institute a new decoration. It would honour the devotion of the 'civilian workers, firemen and salvage workers, and many others who in the face of grave and constant danger' who had won fresh renown for Britain. It would rank next to the Victoria Cross but bear his name: George.

Although intended to reward civilian bravery, the George Cross soon became synonymous with a new form of military gallantry: bomb disposal. This was only natural at a moment when all thoughts were dominated by the Luftwaffe's attacks and their aftermath. What is perhaps surprising is that this brand of heroism should not have remained one of the abiding images of 1940.

The other battles of that autumn and winter are commemorated in phrases that still resonate: 'The Few', 'Finest Hour', 'London can take it', indeed 'The Blitz' itself. Yet once the threat posed by the bombers, and especially by the thousands of time bombs they dropped, had been mastered, the skill and self-sacrifice of the disposal squads began to be forgotten, overshadowed by more obviously glamorous and aggressive feats of arms.

The reasons for this are unclear. It may have owed something to the secrecy in which bomb disposal (BD) work was conducted. No doubt, too, a weary but triumphant nation did not, by 1945, care very much to remember the long months after Dunkirk when it had felt so vulnerable to invasion.

What is certain is that the under-appreciation of BD's worth stemmed from the indifference of much of the military and bureaucratic hierarchy that had created and controlled it. The traditional British scorn for the new meant that for perilously long BD was treated as the Cinderella of the Forces, starved of resources and accorded little respect. Blind to the scale of the threat that would eventually be posed to Britain's ports and cities by some fifty thousand unexploded bombs, the Army's original plan was to man the squads with retired soldiers and those unfit for ordinary duties.

Similarly, its dependence on scientific and technical knowledge meant that BD was regarded by many traditionalists as not being proper soldiering. At first marginalised, Britain's engineers and scientists would ultimately prove themselves essential to a war waged with high technology, but not before BD's scientists had been forced to fight turf wars with the Services' in-house advisers, military men jealous of their role and contemptuous of outsiders.

Between each of the three forces there was also a short-sighted lack of cooperation. Until late in the war, for example, RAF commanders refused for security reasons to tell Army and Royal Navy teams how to make safe unexploded British bombs, even though they repeatedly caused casualties among the advancing troops in North Africa, and later in Italy and France. Indeed, some top civil servants thought that the risks involved in bomb disposal work were to be accepted as no more than routine. For them, the courage shown by BD teams could hardly ever match the standard they thought was required for the highest bravery awards.

Those who served in Bomb Disposal were not the kind to ask for special pleading. Yet they were, and continue to be, short-changed in terms of public acknowledgment of the debt owed them. Though tasked for the six years of the war with protecting Britain, and having almost uniquely been permanently on duty all that while, they were not given their own campaign medal, merely those decorations awarded to everyone who had been in the Forces. Nor is there a national memorial to the 750 men and women of all three services killed on BD work, more than half as many again as those British dead commemorated on the monument to The Few in London.

It would be only invidious to make comparisons between those who helped save Britain first in the air and then on the ground in 1940, but their respective achievements are worth juxtaposing. Fighter Command won its battle with the advantage of a long-prepared strategy, near-parity of numbers with its foe and a vital secret weapon in radar. Bomb Disposal had to contend with an enemy of which initially it knew nothing, which at first overwhelmed it and which was defeated in large measure by sheer courage. At stake in both contests was not simply the British war machine's continued existence but also its ability to prevent Germany from making an unanswerable demonstration of its might on British soil. And the ultimate victory of both arms was vital to the resilience of the nation's morale during the two-year-long night between Dunkirk and Alamein.

The nature of the current wars in Iraq and Afghanistan means that awareness of the lethal properties of explosive devices is higher than it has been for many years. Car bombs, booby traps, IEDs and suicide bombers have become a part of our everyday vocabulary. Yet, memories of the seventies television series *Danger UXB* notwithstanding, few people know much about the early and yet more risky days of BD, which still form the basis of contemporary bomb disposal work. Nor have historians written about it recently for a general readership. My hope is that this book will be a first step in giving those pioneers their due.

Perhaps what sets apart the story of the Blitz and Bomb Disposal from other experiences of the war is its psychological aspect. The Second World War was the first in which extensive bombing was deployed against civilians, and it was bombing above all which made them appreciate that they too were combatants. The Home Front became a battlefront, but more unsettling still was the knowledge that even after the all-clear hidden bombs might go off at any time for days to come. This was to be the first war in which danger was ever present, in which there would be no respite from it, and the effect of that was to strain still further the nation's nerves.

Nor did the men of Bomb Disposal find the demands made of them any less out of the ordinary. For those who had already seen combat, there had at least been an enemy to shoot at and anger to stir the blood. This was a different form of warfare, fought as much with the mind as with the body. Officers had to tackle bombs alone rather than surrounded by their men. Squads would dig down to bombs knowing that they could go off at any second. Life and death were often determined by good fortune or a moment's inattention.

'A bomb is a pretty lethal thing,' one veteran said of the mental pressures generated by the work. 'It is sinister, it is cold, it is still, it doesn't move, nothing, but it might send you to eternity. It's very difficult for people to understand what it's like to be down a hole, a deep hole with a sinister-looking cold steel bomb, that one minute you are there and the next minute it's oblivion . . . We were on our own, and to this day we are still on our own.'

Yet if there was a marked feeling within BD that their achievements were neglected after the war, while it raged their dedication forged a special connection with the population under whose eyes they worked. Voluntarily exposing themselves to great risks, the BD teams helped in no small measure to create the mutual trust between Britain's civilians and its military that would become the keynote of the war, and central to its outcome.

This was the bond expressed in Winston Churchill's exhortations to the nation and his tributes to the bravery of its Armed Forces. When he came to write his history of the war, he made special mention of that exhibited by BD, the importance and dangers of whose work he had been among the first to appreciate: 'Somehow or other their faces seemed different from those of ordinary men, however brave and faithful. They were gaunt, they were haggard, their faces had a bluish look, with bright gleaming eyes and exceptional compression of the lips; withal a perfect demeanour. In writing about our hard times we are apt to overuse the word "grim". It should have been reserved for the UXB Disposal Squads.'

Everyone will have their own ideas as to why the Second World War continues to fascinate us, perhaps more than any other conflict. Historians might say that the reasons for this include an abundance of vivid first-hand accounts, its proximity to our own times, the relatively clear-cut and irreconcilable morality of the protagonists, and its epic scale. It seems to me also to encapsulate fundamental choices that face us all, and to ask us whether we would have made the right ones. Everyone fantasises about being a hero, about being given the chance to save the world: these were the extraordinary men who really did.

1

DANGER UXB

Dusk was coming on and all day the chill air had seeped into their bones. At his listening post Staff-Serjeant Charlie Roberts felt cramped and weary and in need of a change. Rising stiffly, he called to Staff-Serjeant Fox to take a turn with the headphones and walked through the mist to where the other men were taking the rope's strain. There was nothing about this bomb that seemed more than usually dangerous.

It was a Friday in December 1940 – Friday 13th, as it happened, and in the way of soldiers there had been jokes about it being just their luck and about not seeing Christmas. But 64 Bomb Disposal Section were already veterans of this novel branch of warfare. Three months earlier none of them had seen an unexploded German bomb nor grasped the potential for disruption caused by a ton of explosive buried in a station concourse or beneath a giant gasometer. Now the letters UXB – for 'Unexploded Bomb' – were as familiar a sight in London as red buses, and the section had disposed of almost fifty of them without mishap. When it was rumoured that their life expectancy was just ten weeks, and when other sections had met with sudden and violent death, it was a comfort to feel that they knew their work well.

It was a comfort, too, to have Captain Max Blaney back on the job. The thirty-year-old Irishman had kept their spirits high as their Commanding Officer (CO) during those testing months since the Blitz had started in September. His fellow officers thought Blaney almost impossibly debonair, his pedigree as a university champion boxer more apparent than his profession as a civil engineer. Yet his men had appreciated quickly that, for all his zest, he took no unnecessary risks with their lives. His experience had led him recently to be promoted to a role overseeing the work of several of the sections based in this heavily bombed part of east London. This particular UXB was, he had told his sergeant that morning, 'a sticky one', and he wanted to help his old unit and their new officer deal with it safely. It would also give him the chance to try out the latest equipment, on which the backroom boys had been so hard at work.

The bomb had fallen the previous Sunday evening in Manor Park, close to the main road out of London towards Essex. As at thousands of other sites across the capital, an all-too-familiar drill had been enacted once daylight revealed its presence. Acutely aware that the bomb might explode at any moment, the police and local civil defence volunteers had evacuated every flat, house and shop within a radius of six hundred feet. Beyond that, people were told to stay out of rooms that had windows facing the bomb. Taking a few possessions – a change of clothes, a few blankets piled on the baby's pram – the residents of Romford Road had taken a hasty leave of their homes. They were unsure if they would see them or their contents again, or if their houses would join the fifty thousand smashed and damaged each week by the Luftwaffe. At least they had escaped with their lives, when thirteen thousand Londoners had not. Yet it was cold, and they were not sure where they would sleep that night, or the next, or when the bomb squad could come.

The UXB had also played havoc with traffic headed south, to the docks teeming with the raw materials and food supplies that enabled Britain to stand alone against Germany. Lorries bound for the wharves, or for the ordnance industries at Woolwich, had had to seek other, longer routes, only to find side streets choked with trucks slowly negotiating the rubble of smashed buildings. Each delay meant tyre rubber left vulnerable to bombing in riverside warehouses, or Spitfires grounded for want of engine cowlings.

Highly disruptive though this bomb was to the life of Manor Park, there were dozens of others judged to be more urgent cases for disposal. It was not until five days later that 64 Section was able to start work on the site, probing the hole to find the bomb's track and then digging down to expose it. The excavation proved easy, but the find somewhat unwelcome: a 250-kilogram bomb with not one but two fuzes.* The section's new CO, Lieutenant Richard James, a forty-year-old Welshman, was another seasoned campaigner of the Blitz, but Blaney's offer of a helping hand had still been welcome.

The puzzle that confronted them was what had become in recent weeks the bomb disposal officer's classic dilemma, a form of chess played with high explosive. Of the two fuzes in the UXB, that nearest its nose was a Type 17. A (17), as it was written, was the secret weapon whose unexpected appearance that autumn had so perturbed Winston Churchill that he had made mastering it the highest priority. A (17) was a time bomb.

As Blaney knew, once such a bomb struck the ground a clock within it began to run down for up to eighty hours. What made it so lethal was that it was impossible to tell when during that time it might detonate. This necessitated the abandonment for days on end of the key locations at which the bombs were aimed: oil refineries, telephone exchanges, hospitals, railway junctions, government offices, let alone people's homes. And since September 1940, twelve thousand UXBs had been reported. Not all were (17)s, but until investigated all had to be treated as such. It was as if thousands of suspicious suitcases had been found all over London, and beyond.

---

* 'Fuze' is the spelling used by engineers to distinguish the complex initiating mechanism of a bomb from a simple fuse or delay such as the cord in a stick of dynamite.

The obvious solution to a (17) was to pull out the ticking fuze. As bomb disposal officers had found to their cost, however, these were protected by a booby trap that exploded the bomb if extraction was attempted. And, to stop bomb and fuze being moved somewhere else even if ticking, the German armourers had begun to add a second fuze, the Type 50. Deep within its casing, a tiny ball poised on a wire coil would trigger detonation should it be moved as little as a millimetre. Tests later showed that a tap with a pencil on its steel shell could be enough to set off a bomb armed with a (50). Guarded by this, a (17) could count down still more remorselessly.

The reason that James had summoned Blaney was that the UXB at Manor Park had exactly this combination of fuzes. A few weeks earlier and there would have been little that they could do: the bomb could not be moved, and nothing would stop it from exploding. Yet now the odds had evened a little. Scientific analysis of the (50) fuze had shown that it had a maximum life of about sixty hours. After that, the electrical charge that powered it would have leaked away. So, with it having fallen five days earlier, Blaney could now safely regard the (50) as inert.

Moreover, experience had shown that any bomb that had not exploded after eighty hours – once the (17) clock had run down – would not do so, and so after four days evacuees could be allowed back into their homes. Yet that did not mean that the Manor Park UXB could simply be hauled out of its hole. It was also known that while the bomb's (17) was probably a dud that had failed to start running, it might be set in motion by just such a pull as was needed to bring it above ground. Or worse, the mechanism may have jammed while ticking down. British technicians had found more than a dozen different faults in the clock's wheels and gears that could cause this. With an unusual lack of thoroughness, the German fuze engineers had chosen a graphite-like lubricant that lacked fluidity. The mainsprings, too, were made from a brittle material and wound over an asymmetric axle that strained them overmuch, while the milling often had flaws in it. Yet if the workings failed they tended to do so just a few seconds before the bomb's trigger was freed. The tug of a rope or a blow from a spade might be all that was needed to re-start the clock and detonate the bomb without warning.

Blaney was therefore taking no chances. The (17) must be treated as live, which would give the section the chance to prac-tise using two new pieces of technology. First to be fitted was an electric stethoscope – a great improvement on the instruments previously borrowed from local GP's surgeries – which clamped on to the bomb and amplified in a pair of headphones any tick-ing made by the fuze. Eighty yards of flex allowed S/Sjt Roberts to monitor it behind cover, from where came the assurance, in a cheering Scots accent, that nothing was amiss.

With the earth carefully smoothed away from its bulky form, the bomb was now exposed to the thin winter light. Then an important recent addition to the bomb disposers' armoury – nothing less than a clock-stopper – was placed on it. If, scientists had reasoned, the fuze could not be removed then perhaps the mechanism in a (17) could be halted. Their experiments had aimed at producing a magnetic field so strong that it would pre-vent the steel wheels of the clock from moving, even though the fuze was buried deep inside the bomb. The magnet was generated by 250 turns of insulated copper wire, wound together into a coil a foot wide and energised by a surge of 200 amps from half a dozen large and primitive batteries. The coil was placed around the bomb like a collar, with a saddle to seat it on the casing in line with the axis of the fuze. Its field would create enough friction between the clock's moving parts to stop it – although only as long as the magnet was functioning. In tests the coil had shown a disconcerting tendency to burn out when left connected to the batteries for very long.

Blaney did not intend to leave them on for any longer than needed. One of the main reasons that he had come down to Manor Park was to give some instruction to the men on the use of another newfangled piece of kit. Like many of the inventions devised in haste in those frantic months when invasion seemed imminent, the steam steriliser combined engineering ingenuity with an eccentric appearance. Despite the stereotypes perpetuated by the military, the machines' air of Heath Robinson oddity was rarely of the scientists' doing, for their thinking was usually sound enough, and often inspired. Instead it was largely due to parts being scarce and cobbled together from items intended for other purposes. The heart of the steam steriliser, for example, was a portable but none-too-reliable domestic boiler.

Though it looked a little absurd, the steriliser offered the first real hope of defeating the menace of the (17). The principle behind it was simple enough: if the fuze could not be extracted without setting off the explosive, then maybe the explosive itself could be removed. Then, when the clock ran down, there would be nothing to detonate.

The mechanics were similarly straightforward. A machine automatically cut a hole in the casing of the bomb and inserted a pipe that pumped in a mixture of steam and soap to dissolve the explosive. As it melted, the explosive flowed out down the same pipe. The whole operation was driven by pressure from the boiler, with no need for anyone to approach the UXB once the equipment had been clamped to it. In a war fought for the first time as much with science as with soldiering, such breakthroughs could avert defeat or hasten victory.

Only there was a problem. Blaney had been waiting for the steriliser to arrive but – no doubt because it was Friday 13th – the lorry carrying it had broken down. No matter. Ideally he would have liked to have steamed out the UXB in the crater it had made, so avoiding the risk of moving it. Yet with the clock-stopper in place there should be no danger. It could always be sterilised somewhere else tomorrow. That would serve just as well as a demonstration.

Moving quickly, before it got too dark, the squad fitted the thick coil of the clock-stopper around the belly of the bomb. Then a block and tackle was slung over the joist of a neighbouring house, only for it to become apparent that the coil was in the way of the lifting strop for the hoist. The only course of action would be to remove the clock-stopper for the brief period needed to raise the bomb, then re-attach it immediately it reached the surface.

Roberts, whose usual role was to work the steriliser, asked his fellow staff-serjeant Fox to relieve him of the headphones but to keep listening in case the (17) began to run. This was a rare chance for Roberts to get his hands dirty, and he joined the line of men hauling on the rope. Up came the bulk of the bomb until it was suspended a couple of feet above the hole. It began to swing a touch freely and Blaney stepped forward to cushion it with his hands. As he did so, it exploded.

In the evidence that he later gave to a military inquiry, Fox noted that the bomb could not have been in a more lethal position. In a microsecond it was transformed into a searing blast of noise, light and heat that tore outwards at supersonic speed, clawing at flesh, pulverising bone, annihilating all in its path. Max Blaney and Richard James died instantly, as did Charlie Roberts, Lance-Corporals Doug Mills and Stan White, Sappers Joe Maycock and Ted McLaren, and Drivers John Pickering and John Lauchlan. Most of the men were in their twenties. Also killed was Inspector Henry Lane of the Metropolitan Police, who had been watching the operation from across the road. The next day a uniform button was found embedded deep in a door frame away down the street. Very little else that was found was so easily recognisable.

Fox survived through sheer luck, having seconds before changed places with Roberts. He told the inquiry, too, that he was certain that he had heard no ticking from the (17) fuze. Blaney had taken all the recommended precautions, so something had gone terribly wrong. Until the reason for that was established no other BD squad could feel safe. Had it been the result of a new modification to the (17) which made it tick silently? It was vital to discover – and quickly – why the (17) had detonated.

The unexpected answer was that it had not. It was not the (17) that had exploded the bomb but the other fuze, the (50) that Blaney had thought was not a threat. BD's research scientists had for some time suspected that their German counterparts had been improving the electrical storage capacity of the fuzes, so enabling them to remain live for longer. The British experts then thought of something else that had been overlooked.

It was known that fuzes could be discharged by heat: if the bomb was made hot, the electrical charge leaked away more quickly. But why should the reverse not also be true? That is to say, what would happen to a (50) if it was kept somewhere unusually cold – such as in a hole in the ground, in London in mid-December? In time, experiments would show that a (50) under such conditions might remain sensitive not for sixty hours but for a thousand. That knowledge would have been the difference between life and death for 64 BD.

Max Blaney was only one of twenty Bomb Disposal personnel killed that week. Since the start of the Blitz, more than two hundred of his comrades had died or had been maimed in dealing with bombs that had fallen but had failed to explode. Almost nine thousand of those bombs had been discovered just in the month before Blaney's death, and many other men would lose their lives in the weeks that followed as the need for urgency and for information about UXBs led to the taking of extreme risks.

The incident that destroyed 64 BD Section had a particularly high toll, but it is representative of the main features of the war against the bombs. It was to be a lonely and often frustrating struggle, one that sometimes appeared less a part of a wider war than a private battle between the German fuze engineers and the British bomb disposal officers and scientists. No sooner had one type of fuze been worked out than a new variation seemed to be adopted, and for long periods the Germans' remarkable capacity for practical invention gave them the upper hand.

If it was the courage of the BD officers that initially evened the odds, it was the ingenuity of the scientists that in the end allowed the squads to prevail. It was a pattern of development that mirrored that of the British effort as a whole. Its early days were marked by muddle and improvisation in which amateurs responded bravely to challenges for which they were wholly unprepared. In time, and at the cost of hard lessons, BD was to transform itself into a highly professional organisation that won out not by chance, nor simply through larger resources, but by better planning, superior execution, greater endurance and smarter thinking. Germany had begun the war as the masters of bomb technology, but by the end Britain had become unrivalled at countering it.

2

THE SHAPE OF THINGS TO COME

One warm afternoon in July 1849 the whole population of Venice stopped what it was doing and in silence looked at the skies to the west. For months they had been blockaded by the Austrian forces, from whose rule they were trying to free themselves. So far the Venetians had held out, but drifting towards them on a south-easterly breeze was the future of warfare.

Outlined against the summer sky were dozens of small balloons that had been released from the Habsburg fleet moored in the lagoon. Dangling from these were explosives fitted with slow-burning fuses. The Austrians were making the first attempt in history at bombardment from the air.

Events were to show that the new strategy still needed some refinement. Accompanied by cries of '*Buon appetito!*' from the Venetians, the balloons were carried by the wind away from the city and over the lines of the besieging army on the mainland. Nevertheless, if the technology could be better harnessed, the possibilities were self-evident.

Explosive devices have a long history. As early as the tenth century, the Chinese had developed powder-fuelled rockets, and by the Middle Ages artillery was an established feature of Europe's battlefields. Yet attacks launched from the air had to wait another five hundred years, and for the discovery of flight.

The novel form of war introduced at Venice was banned by the first Hague Convention, signed in 1899. The prohibition was then rescinded by the second, eight years later. In the interval, the aeroplane had been invented and the military potential for aerial bombing foreseen. As with landmines in more recent times, self-interest prevailed over moral posturing.

Having been the intended victims of that early assault, the Italians were now to be in the vanguard of waging war from the air. In 1911 they claimed the doubtful honour of making the first successful attack from an aircraft when they bombed Ottoman troops at Tripoli. Their missiles were fragile containers filled with nitroglycerine that exploded on contact with the ground: 'a wildly hazardous procedure', wrote one historian of ordnance, 'well in keeping with what has since become part of the Italian military tradition'. By the start of the First World War another nation had taken a clear lead in developing this new technology: Germany. Paris was bombed in 1914, and attacks mounted as far afield as Russia and Greece. In 1915, the first air raids against the United Kingdom were made by Gotha heavy bombers and Zeppelin airships.

More than 1400 people were killed by the 8600 bombs dropped on Britain during the next four years, and another 3400 were injured by them. Although some of the casualties were caused by indiscriminate bombing, most of the time the intended targets were military and industrial. More than fifty attacks were made by Zeppelins, and many aimed at production centres such as Sheffield, Liverpool and Manchester. The London Docks were hit in September 1915 and August 1916, and more than a dozen soldiers killed at an Army camp near Shorncliffe in October 1915. Although the raids were regarded by some as a perversion of science, and stoked propaganda about airborne 'baby killers', the British were not slow to reply in kind; one of their first strikes, on

Christmas Day 1914, was directed at the Zeppelin hangars at Cuxhaven, in the German Bight.

By that war's end bombs had assumed many of their modern characteristics. Both sides had developed incendiary as well as high-explosive armaments, while the Germans had started to manufacture bombs in steel rather than in iron and cast with a form that took account of aerodynamic principles. Great advances had been made, but by comparison with what was to come in the next twenty years, bombing was still relatively primitive. The largest German weapon weighed a thousand kilograms (2200 lb), but an aircraft was only able to carry one of them; by contrast, the payload of a Second World War Lancaster bomber would be ten times greater.

Similarly, fuzes had not really progressed beyond simple arming devices. The fuze is the brain of a bomb, the animating force of what is otherwise only so much muscle. Mining engineers and soldiers had long been able to control the timing of explosions with simple fuses – a trail of gunpowder or a length of cord known to burn at a fixed rate. As early as the American Civil War, artillery shells were fitted with short-delay fuses to ensure that they did not explode before reaching their target. The addition of mechanical components during the First World War made the fuzes – so spelled to indicate their greater complexity – of aerial bombs more sophisticated, but in essence they were still basic detonators that set off their charge on impact. An estimated 6 per cent of German bombs up to 1918 failed to explode because of some malfunction. To make these five hundred or so safe, the Royal Army Ordnance Corps needed to do little more than gather them up like autumn leaves. That, however, was about to change.

After Germany's defeat, the terms of the Treaty of Versailles hobbled the once-mighty Teutonic arms industry. One of the few large firms not affected was Rheinische Metallwaren- und Maschinenfabrik Aktiengesellschaft, founded in 1889 and soon known as Rheinmetall. Based in Düsseldorf, it had permission under the Treaty – unlike its rival Krupps – to continue armaments research, and in 1926 it advertised a vacancy for an electrical engineer. The post went to an energetic twenty-four-year-old from Leipzig named Herbert Rühlemann.

Rühlemann inherited his intellectual curiosity from his mother, a teacher, who came from a more bourgeois background than her train-conductor husband, and who encouraged her son's scientific ambitions. Another important influence was an older brother-in-law, a dealer in the new wonder material aluminium, whose success in business allowed him to acquire such luxuries as a telephone. Excited by the possibilities of the modern world, Herbert trained as an engineer and then gained early experience in local laboratories and factories.

His chance of self-advancement came through Germany's growing fascination with another new invention, the wireless. Rühlemann became one of his country's few experts on broadcasting equipment and soon his prospects were such that he felt able to get married, to Sophie Reuschle, a well-known children's writer some years his senior. Yet, despite his involvement on a daily basis with the leading edge of commercial science, so far his life and outlook had been firmly anchored in provincial Germany. That looked unlikely to change when he took the job with Rheinmetall and moved to its plant at Sömmerda, near Erfurt in placid Thuringia. Within a decade, however, his work would be at the heart of the risen Reich's plans for Europe-wide domination.

It was clear to Rheinmetall that aircraft would be increasingly important in war, and experience in the trenches had shown that it was usually beyond the capacity of contemporary artillery to bring down such a swiftly moving target. One solution that now occurred to its engineers was to fit shells with a timed fuze that would explode them automatically near an aeroplane, obviating the need for them actually to hit it. Proximity shells already existed for conventional field guns, but Rheinmetall's problem was that other firms held the key patents. If Rheinmetall was to develop an anti-aircraft shell, it would need to devise fuzes of its own. Advances in technology suggested that electricity might prove one way of triggering such weaponry, but Rheinemetall's fuze departments, like those of all the arms firms, was mechanically orientated. So, for expertise on electricity they turned to Rühlemann.

High explosive (HE) is more stable than its name suggests. The British Army has been known in recent years to use it – unofficially – as a quick-burning cooking fuel in the field. High explosive needs to be relatively inert so that it can be safely transported and stored in bulk. Simply holding a match to it will not set it off. Rather, in a bomb its potential is released through a carefully built-up series of ever more powerful explosions.

This process is rather like lighting a coal fire, as the explosive train graduates from matches to newspaper, then to kindling and finally the fuel itself. Each material is progressively less flammable but burns more vigorously once alight. In a bomb, the matches equate to a percussion cap, which then triggers a detonator, the booster charges and then the HE.

This succession of events happens very rapidly. In about 1/40,000th of a second the explosive is prompted to change its state to what is – scientifically at least – a more stable form, usually gaseous. During that change, energy is violently released as heat (at about 3–4000° C), sound, light and pressure. What makes a bomb so dangerous is the vast volume of superheated gas produced – up to fifteen thousand times the original mass of the explosive. The impact of the blast on anyone within its radius can be, according to one of the Army's senior instructors, 'rather worse than being crushed against a wall by a juggernaut doing a hundred miles an hour'.

It is the fuze that releases this explosive potential, and what Rühlemann was asked to research was whether electrical energy could be used to trigger it instead of the standard mechanical components. If so, fuzes could be made not just more reliable but also more versatile, for if the moment of a bomb's arming could be delayed until it was nearer or even within the target then the effect of its explosive charge would be rendered more precise.

Rühlemann, however, was to find himself largely on his own at Rheinmetall. In his memoirs he complains frequently that none of his mechanically-minded bosses at the firm understood his work, nor did he have the resources he needed. When he wanted to buy a voltmeter for his laboratory, his request was rejected by managers who told him instead to go to the assembly department and use its *boulanger* – an instrument which measures muzzle velocity, and entirely unsuited to his purpose.

Despite these handicaps, within six months he had made what would prove to be a breakthrough in fuze design. Working at first with artillery fuzes, he suggested that they might be detonated by electrical igniters that drew on a reservoir of stored energy.

Basic batteries did exist, but Rühlemann rejected these as having too limited a storage life for his needs. Instead he concentrated his attention on condensers. The descendants of the Leyden jars devised by German and Dutch physicists in the eighteenth century, condensers – or capacitors, as they are also known – consist of a pair of conductors kept apart by a narrow insulator. When a voltage is passed across the conductors an electric field is created in the insulator, which can retain energy.

Rühlemann's inspiration was to appreciate that the rate at which this energy leaked through a series of condensers could be used to control the timing of an explosion. Proving this, however, would require expensive experiments and Rühlemann realised that if he was to secure funding he would have to make progress quickly. He therefore turned his attentions from artillery fuzes to impact-detonated bomb fuzes as being the easiest to develop in a short time.

Nonetheless, manufacturing even these would present a major challenge. As he recalled afterwards, 'none of the major four components (the condenser, resistor, cold cathode diode and igniter) as on the market was adequate for the purpose'. Unable to get

sufficient help from Rheinmetall, Rühlemann turned instead to
the colossus of the German electrics industry – Siemens. There,
he felt, his research would be given the respect it merited.

The priority was to make condensers that would offer a large
storage capacity in a small space and had a high resistance. From
sheets of thin metal foil separated by strips of varnished paper,
Siemens produced condensers that had resistance a million times
higher than any previous model. They also made the resistors –
porcelain cylinders sprayed with compressed carbon – and entirely
new high-precision diodes.

The other key component was the igniter. Rühlemann knew
that electrically-operated igniters had been used in the mining
industry for some time. Indeed, that fact had made him think that
an electric fuze was itself a possibility. Yet all the igniters he tried
proved too insensitive, needing at least 1000 volts to function, far
more than his small fuze would be able to supply. The only way
to lower the required voltage would be to find a wire much finer
than any customarily used in igniters. Showing great persistence
and resourcefulness, he eventually discovered near Cologne a small
firm that specialised in working with a chromium–nickel alloy.
Pound for pound, this cost more than gold, but a single gram of
it could be spun into a wire two miles long – enough for a million igniters.

By the summer of 1929 Rühlemann was able to begin field
tests, and although there were to be many small modifications
during the next decade, the essential form and design of the electric fuze had already been established. Made first of brass and
later of aluminium or steel, they were cylindrical in shape, some
four and a half inches long and about half as wide. The fuzes were
generally divided into two sections, with the resistors and capacitors in the bottom part and the switches in the top.

One other development was their position in the bomb itself.
'Our design evaluation,' wrote Rühlemann, 'had shown that the
most logical place to locate the electrical bomb fuze was not at the
nose of the bomb but somewhere at the middle, within a tubing

perpendicular to the axis. This location would simplify our charging equipment.'

Electric fuzes offered many advantages over mechanical ones. They would be more reliable, having fewer moving parts to jam or to wear down. They would also eventually allow for more accurate setting of delays. But above all they made bombs safer to transport.

Handling munitions is inherently risky. The largest explosion on British soil happened at RAF Fauld in Staffordshire in 1944, when almost four thousand tons of ordnance was set off accidentally. The cause was probably a spark from a chisel that was being used to remove a detonator. Seventy-five people were killed as the depot and a nearby factory vanished into a crater three-quarters of a mile across. Such dangers were ever present at airfields and on aircraft, as bombs were frequently loaded live or fuzed when the aeroplane was in flight. This could lead to premature explosions or create problems if a fault forced a bomber to return with its payload still on board.

Working to the requirements of the German military leadership Rühlemann's team had to devise a method of passing current into the fuze and its condensers only *after* the bombs had left the aircraft. Their solution was a telescopic arm that attached to a charging cap held on to the boss of the fuze by a spring. As the bomb was released it pulled the arm down. This caused two sliding contacts to meet and a current of 120 volts to flow through leads in the arm to the fuze boss and then into the reservoir condenser. A moment later the charging cap – which remained connected to the arm – was plucked from the fuze as it continued to fall.

While the bomb was dropping towards its target the energy now stored in the reservoir began to pass into a firing condenser with a high resistance. This took about six seconds, giving the aircraft time to escape. The great advantage, of course, was that since the fuzes did not become armed until after they had been released, if there was a hitch earlier in the mission then the bombs could safely be jettisoned or brought back to base.

Mounted at right angles to the axis of the fuze was a fine wire on which balanced two small steel balls. On impact with the target this vibratory switch made contact with a metal ring around it, causing the firing condenser to discharge through two thin strips of metal known as the firing bridges. This then set in train the detonation of the explosive.

This clutch of technological advances was far ahead of anything being done by the British. When the Second World War began, the RAF's bombs and fuzes were little different from those used by the Royal Flying Corps in 1918. During the twenties and much of the thirties, the unwillingness of successive governments to contemplate the possibility of another European war meant that there was little money or encouragement for ordnance research. When war came, the RAF's stock bomb – the five-hundred-pound General Purpose – lacked the strength to penetrate hard targets intact and had insufficient explosive to do much damage.

The standard British fuzes, which were known as pistols, were purely mechanical. Again, development of these during the inter-war years was negligible and they operated in the same manner as some German fuzes had in 1918. This took the form of a safety device consisting of projecting radial vanes that rotated under air pressure as the missile fell. After half a dozen revolutions, a spindle disengaged from a striker and the fuze was armed. The British consistently had trouble with their pistols into the early years of the Second World War. As they needed to be assembled on the ground, before the bombs were loaded, there was more potential for things to go wrong. The modern world also revealed their lack of sophistication. For example, as aircraft became faster the bombs they released fell at a higher speed, causing the vanes to rotate more rapidly and fuzes to arm dangerously close to their bombers.

Supported by the revitalised German Armed Forces, Rühlemann was able to conduct extensive tests to eliminate any such technical problems. There was much to be done. 'Bombs dropped from aeroplanes,' he wrote, 'were a new kind of weapon.

Not much was known of their ballistics, their accuracy in hitting a target, their maximum velocity and other parameters.' Trials would also have to be undertaken to determine details such as arming times and the sensitivity of the impact switch. Yet all of this would have to be done in secret.

One of the prohibitions placed on Germany by the Versailles Treaty was the production of weaponry for export. When Rhenimetall received an order for ammunition from China it got around the rules – Rühlemann later confessed – by having it made under licence in Italy. Tests on weaponry were also banned, but by the mid-twenties Germany's military and industrial leaders were collaborating on secret evaluations.

Thus the work on Rühlemann's charging gear was done covertly at the Heinkel factory at Warnemünde on the Baltic. Artillery tests were used to disguise the noise of work on bombs and aircraft, although Rheinmetall used this to their advantage, converting a field gun on Lüneburg Heath to mimic the effects of dropping missiles from the air. It fired bombs into concrete roofs and ship decking to check that both casing and fuzes would survive the impact. When the Sudetenland was appropriated by Germany in 1938 the practice targets became the Czechoslovaks' modern bunkers and fortifications.

The central figure in this testing process was General (Engineer) Eric Marquard. In 1933 he was given command of Abteilung LC7 (Bomben), the newly created department tasked with developing air-dropped weaponry, but his relationship with Rühlemann had started some time earlier. Owing to the ban on re-armament, at first its existence had to be concealed and Rühlemann's letters to him were addressed to a false name.

Together, Rühlemann and Marquard had spent the summer of 1932 testing the electric fuze in the most extraordinary circumstances. 'Secrecy was the key word for the entire operation,' recalled the former, who had told his wife that he was on a naval ship in the Atlantic. In fact, Rühlemann, Marquard and a flight of German bombers were deep in Soviet Russia.

Since 1924, in return for examples of the weaponry being assessed, the Russians had offered the still-clandestine German Air Force secret use of a proofing ground at remote Lipetsk, halfway between Moscow and Volgograd. There the Germans harvested the fruits of their illegal re-armament – well advanced even before Hitler took power – without arousing the world's suspicion.

The great success of the summer was Rühlemann's brainchild. 'We had not one failure with electric fuzes during all our tests,' he wrote afterwards. 'For me it was the break-through point.' The airmen were most pleased with the flexibility offered by the standard fuze, the EL25. 'It could ignite the bomb without delay, at the moment the bomb touched the target. But by selecting a different charge in the airplane, the same fuze would ignite the bomb with a .003 [second] delay, after the bomb had penetrated into the target. Further, a long delay of 12 seconds could be selected to make the fuze usable in dive bombing applications.'

The German Navy, which had followed Rühlemann's research with interest, now agreed to back his work, and then in 1935 he received his first large order. It was from Hermann Goering, commander of the Luftwaffe, which he had recently established in flagrant violation of the Versailles accord. For his bombs, he told Rühlemann, he wanted two hundred thousand electric fuzes.

Although Rühlemann was collaborating closely with the Luftwaffe, Rheinmetall was nevertheless a business and one that aspired to sell its products as widely as possible. Following the demonstration of the fuzes' effectiveness by German pilots in 1936, during the Spanish Civil War, Rühlemann made a sales trip to an air show in England. Then, in the spring of 1938, he travelled to America to show the EL25 to the US military. He also visited leading firms such as General Electric and Westinghouse, but though impressed by the American capacity for mass production he thought their condensers inferior to his.

Perhaps naively, he was surprised by the critical tone of articles about the German government in the American press. Rühlemann states in his memoirs that he was not a supporter of

Hitler before he became Chancellor, but saw that he managed to solve the country's most pressing problem, mass unemployment. Like many other well-educated Germans, Rühlemann effectively shut his eyes to the implications of the Nazis' policies.

'I was at no time in a position to have a significant influence on the events which occurred within this century,' he avers in his autobiography, *Father Tells Daughter*. 'Discussing them and their influence on my life will be objective, because I have no reason to defend or condemn them.' Certainly, he was no Nazi zealot, but his claim to be a neutral observer of his own behaviour was at best disingenuous, not least because his work for the Luftwaffe had made him rich. On his return from America Rühlemann was summoned by Goering and asked for his impressions of the country. He was then given a present of fifty thousand Reichsmarks (now £225,000) and, a little while later, told to join the Nazi Party.

By 1939 Rheinmetall had taken out more than eighty-five patents on a range of electric fuzes. Thirty-eight million Reichsmarks had also been sunk into construction of a vast new factory dedicated to making fuzes at Breslau.

To complement the fuzes, the Luftwaffe had developed powerful new explosives and a specialised assortment of bombs, graded according to the target's size and characteristics. Informally, armourers gave the bombs names that reflected their progressively less svelte and more menacing silhouettes: Dolly, Diana, Lisa, Hermann, Fritz and Satan.

Where the Germans' preparations for the coming bomber war were characteristically inventive and thorough, those of the British to resist it were for too long muddled and meagre. There was reluctance in the country at large to face up to the inevitability of another conflict, while almost until the advent of war government departments tasked with contemplating its consequences passed the buck.

Yet in fairness to those charged with making arrangements for Britain's civil defence, it must be acknowledged that they had little reliable information on which to base their plans. The only certainty about a future war seemed to be wide agreement with Stanley Baldwin's gloomy prognosis that the bomber would always get through – wave after wave of aircraft streaming over the English coast as envisioned by H. G. Wells in *The Shape of Things to Come*.

In reality, believed one scientist later involved in building shelters, 'The starting point of any account of the situation with regard to air warfare or its consequences for the civilian population, as they appeared in 1939, must be to emphasize the total ignorance of everybody and everything to do with the subject.'

Those thought to be experts, such as the Air Staff of the RAF, told the Government to think in apocalyptic terms. The first projections of civilian casualties made in the twenties were relatively low, based as they were on data from Zeppelin raids. The forecasts climbed steeply a decade later, however, following the devastation wrought from the air on Spanish cities including Guernica and Barcelona. By June 1937 the calculation was that sixty-six thousand people would die every week in any aerial campaign waged against Britain, and in the spring of 1939 the Ministry of Health was assuming that 1.8 million would be killed or wounded in the first six months of enemy bombardment.

In the event, these figures proved gross overestimates: about fifty-five thousand people were killed or injured in the first six months of intensive bombing. Perhaps the main cause of this miscalculation was that the attacks in Spain – the scale of which had been in any case exaggerated by supporters of the Republican cause for propaganda purposes – took place against a populace out in the streets during the day, while those of the Blitz were to be largely against a people under cover at night-time. Nevertheless, the dire predictions meant that much work was done in the late thirties, albeit often patchily and hurriedly, on the construction of shelters, on the feasibility of evacuating London's children and on familiarising all with air raid precautions. Another abiding preoccupation was the possibility – again unrealised – of an attack with gas. Only on the day after war broke out, however, was the Ministry of Home Security established to coordinate civil defence schemes hitherto left unsatisfactorily to local government. Even then, uncertainty about its remit and its evident lack of clout in Cabinet led it to be swiftly nicknamed 'The Ministry of Some Obscurity'.

If the civil authorities tended, for lack of facts, to inflate the scale of destruction with which they would have to deal, the reaction of their military counterparts to the bombing threat bordered, for much the same reason, on the unimaginative. Again, this was perhaps because not until war came were they were able, indeed forced, to appreciate the advances that Rheinmetall had made. Only hard experience could reveal the real emergencies created by mass bombardment.

Although the Army had made safe German UXBs during the First World War, it was the police who thereafter had dealt with the most common threat posed by explosives, the sporadic attacks of the IRA. These continued as late as June 1939 – when seventeen people were hurt by bombs in Central London – and even into the early months of the war itself, but the devices were rudimentary. They usually consisted of sticks of gelignite, which could be disarmed by having wires cut or, as seems from newspaper reports to have been standard practice, by being plunged by alert constables into buckets of water.

As war neared, however, there was a growing awareness at the War Office that better provision should be made for dealing with any enemy bombs that fell on Britain but failed to explode. Yet the volume of these, and the chaos they would create, could not at this stage be foreseen, given the limited evidence from Spain. Instead, the chief concern of the War Office was to find another department on which to lay this burden.

Conferences, wrangling and compromises continued throughout 1939. It was envisaged initially that the Home Office would accept the responsibility for UXB work, with the Army training a designated force possibly drawn from the retired soldiers of the

British Legion. Then it was thought that perhaps ARP wardens could take it on, following a day's course on bombs, and until they were ready the War Office agreed to place parties of Royal Engineers (RE) near important targets. It was next proposed that these could instruct civilian volunteers, but few appeared and some local authorities – already bearing the expense of building public shelters – jibbed at taking responsibility for UXBs in their area.

In all of this it was assumed that any unexploded bombs presented a passive rather than an active threat. They would be found on or near the surface and would need little more than a wall of sandbags and a small demolition charge to tidy them up. This was probably why the task was delegated, until a solution was found, to the Royal Engineers, much of whose regular soldiering involved explosives and the building of fortifications.

So it was that in November 1939 the first bomb disposal parties were formed. They consisted of just three men, a junior NCO and two sappers, equipped with seventy-five pounds of gelignite, two shovels, two picks and five hundred sandbags. No information was available on the types of bombs with which they might be confronted, nor on the workings of their fuzes, and the only instruction they were given was in how to place sandbags around a shell prior to demolition, as if on a firing range.

Among the early recruits to these units was nineteen-year-old Harry Beckingham, a draughtsman fresh out of technical college. He was given a day's training in Sheffield, 'at the end of which we were given a drawing which showed how to deal with an unexploded bomb'. This depicted a wall being built around the bomb with corrugated metal and bags of sand, with room left to crawl in and place a charge.

In some haste, these small parties were dispatched to locations thought to be especially vulnerable to bombing. Yet the widely expected raids failed to materialise, and as the months passed the problem of bomb disposal seemed less and less urgent. Belgium, the Netherlands and France still barred the way for the Luftwaffe, which struck only at shipping in the North Sea and at the remoter coasts of Scotland. With little work for the bomb-disposal trios to do, Beckingham later wrote, 'we languished in idle obscurity waiting for something to happen'. Cut off from their parent companies, they became odd-job men for whichever unit was nearest. Britain appeared convinced that it was immune to attack, and the Army that bomb disposal was a flap about nothing. Beckingham spent most of his time at the YMCA, playing table tennis.

Nevertheless, some progress had already been made in fathoming the technical secrets of German fuzes. After the war, a myth grew up in Bomb Disposal that much time and many lives might have been saved if only someone had remembered that Rheinmetall had registered their invention with the Patent Office in London in the late 1930s. Instead, so the story went, the diagrams were not found there until 1945.

In fact, British ordnance engineers were well aware of Rühlemann's work almost as soon as it came off the drawing board. As early as March 1936 an example of the new electric fuze was being studied by the Royal Aircraft Establishment, the aeronautical research centre at Farnborough. Although the origin of their specimen is uncertain, since this was before its use in the Spanish Civil War, its novel aspects were clearly enough understood. 'The scheme is ingenious,' ran the report, 'and has the advantage that there is no source of electrical energy inside the bomb until the bomb is actually leaving the aircraft.'

The device was demonstrated to the Fuze Committee at the Air Ministry. Its assessment seems not to have been shared with its military and naval counterparts, nor was much more interest taken in electric fuzes by the RAF until the eve of war. Then, one evening in the summer of 1939, an airman, Len Harrison, found himself presented with a heap of bits and pieces with German markings.

Harrison had spent most of his adult life in the Air Force. He had joined as a boy apprentice at sixteen, qualified as an armourer and seen service with the Fleet Air Arm on the China Station before retiring in his late twenties to his native Devon. Within a few years he was back in the Reserve, and now, at thirty-three, was the senior civilian instructor at the RAF's newly built No. 1 Air Armament School at Manby, Lincolnshire.

The squadron leader who had brought the scraps of metal up from London told Harrison that it was the remnants of a German electric fuze. 'I couldn't argue with this,' said Harrison, 'because frankly I didn't know what a German fuze looked like.' He was ordered to re-assemble it and forbidden to tell anyone what he was doing. Not being an electrician, however, he had to take a signals expert into his confidence to help him. 'It was the blind leading the blind,' he recalled, but rather to his surprise, after a fortnight's work in their spare time they had managed to put the fuze together again. Harrison's report, and his suggested method of dealing with such a fuze if found in a UXB, led to a meeting with Wing Commander John Lowe, a specialist in air intelligence and adviser to the Ministry of Home Security (MOHS). The two men were soon to put Harrison's theories to the test.

The first of the 71,000 tons of bombs dropped on Britain during the Second World War fell near the Scapa Flow naval base on 17 October 1939. The first unexploded bombs landed the following month in the Shetlands, having been aimed at the seaplanes stationed at Sullom Voe. It was these that were to become, in December 1939, the subject of the earliest attempt to recover and examine German UXBs.

The mission was organised by the fledgling MOHS which, by virtue of its assumption of responsibility for ARP, became by default the department concerned with unexploded bombs. Its chief adviser was Reginald Stradling, a civil engineer who before the war had run the government's Building Research Station and directed much of the work on bomb shelters. In the early weeks of the war, the key personnel of the new Ministry were moved out to Princes Risborough in Buckinghamshire. There, Stradling studied the use of protective camouflage and the effects of bomb blasts, and helped to develop the Anderson and Morrison shelters.

His choice to lead the Sullom Voe party was the somewhat unlikely seeming figure of the Secretary of the Faculty of Architects and Surveyors. Arthur Merriman was however a technocrat to his dextrous fingertips, one of science's new men who had repaid the Victorians' faith in the power of education to overcome the handicaps of humble origins. The son of a warehouseman in the cotton industry, he had been brought up in Manchester by his grandfather, a baker's clerk. A gift for mathematics won him a place at Cambridge, from where he had graduated in 1915 and, once in uniform Merriman's scientific bent secured him a commission in the Ordnance Corps. He ended the First World War as a captain, having acquired a wide knowledge of fuzes and ammunition.

In the years after the war Merriman combined a career teaching engineering with further study, including a doctorate at Lille University. It was some way from the mills of Lancashire, and evidence perhaps of a drive that brought him to Stradling's attention.

With Merriman was another expert on fuzes, who would leave his mark on the early history of bomb disposal. Squadron Leader Eric Moxey had also seen action in the Great War, initially in khaki with the York and Lancaster Regiment. In 1916 he had survived the decimation of the Sheffield City Battalion on the first day of the Somme, and the next year had transferred to the Royal Flying Corps. A motorcycle enthusiast, Moxey had long been fascinated by speed and machines, and had raced in the Isle of Man TT. Like Merriman, he was now in his late forties with sons of service age, but in early 1939 the prospect of another war prompted him to join the RAF's Volunteer Reserve.

What the pair found in the Shetlands was to greatly expand the understanding of the behaviour of UXBs, which until then had been limited by a lack of examples. The first thing they discovered, rather disconcertingly, was that the Luftwaffe had managed to make twenty-three passes over the base. They had left half a dozen unexploded bombs behind, four of which appeared to be inert and had already been gathered up.

These each weighed fifty kilograms, the smallest size of conventional bomb used by the Germans, and Merriman's examination of one revealed several points of interest. The six-inch-long fuze was held in a hollow pocket that ran across the width of the bomb, making it less susceptible to damage as the missile landed nose first. It was locked in place by a steel retaining ring. When this was unscrewed and the bomb rolled on its side, the fuze simply fell out. Presciently, Merriman suggested that a finger be kept on it during extraction in case a spring-operated booby-trap was ever placed underneath.

Screwed into the base of the fuze cylinder was a stubby steel container known as the gaine. This was filled with penthrite wax, a sensitive explosive. The gaine nestled in the fuze pocket on top of and between several booster pellets of another explosive, picric acid. Together these initiated the chain reaction needed to detonate the main filling of the bomb, which made up about half its weight. When the fuze ignited, a percussion cap inside it would flash into the gaine, setting off the penthrite wax. This in turn would detonate the picric pellets, causing the HE to explode. The entire sequence was virtually instantaneous.

It was comforting to have now filled in some of the blanks about the bombs' mechanism, but one of Merriman's observations seemed likely to herald new problems. Those bombs that had exploded near the base had done little damage beyond killing a rabbit, although they had been powerful enough to hurl fifty-pound rocks some four hundred yards. The UXBs had fallen further away, however, on wet peaty soil. They had penetrated about four and a half feet, and then stones had deflected them sideways and upwards, so they emerged sixteen feet from the point of entry and came to rest a further six feet away. And these were only relatively small bombs. Clearly the notion that UXBs could simply be collected for disposal after plummeting to earth would have to be revised. It looked as if many would bury themselves deep in the ground. There might well be much digging ahead.

Once the fuzes had been dissected in the laboratory a number of their other features emerged. As was already known, the fuze began to be charged as the bomb left the aircraft. It now became clear that the charging system allowed for a variety of arming times, depending on which of two plungers in the fuze head had been selected. The first would cause the bomb to detonate on impact, having taken eight seconds to charge while falling so as to let the aeroplane get clear. The other offered two options: a charging time of eight seconds but with a pyrotechnic delay of 0.3 seconds, enabling the bomb to penetrate a structure before exploding; or a charging time of just two seconds, followed by a delay of 8.8 seconds, corresponding to a low-level attack and time for the pilot to pull away.

As Merriman realised, this method of arming the fuzes created a problem that had not been anticipated. Since it took up to eight seconds to charge the condensers, 'a condition may arise that when dropped from low altitudes (less than three hundred feet), the fuze condensers may not be fully energised at the moment of impact'. In other words, below a certain height the bomb would not explode when it hit the ground, but as the charge continued to leak from the reservoir into the firing condensers it would belatedly arm the fuze. Eventually this energy would discharge itself harmlessly, but in the meantime any other vibration or impact – such as a blow from a pick – might close the circuit and provoke an explosion. It would also be impossible to tell whether any bomb discovered intact had malfunctioned and so presented no risk, or was one that might still go off.

Not only was it going to be more difficult than initially expected to recover UXBs, but it might also prove dangerous trying. When found, one solution might be to blow them up in situ, but evidently there were going to be times when this would not be feasible. Bombs were going to have to be made safe by having their fuzes extracted, and extracted safely.

The recognition of this led Merriman and Lowe to devise what became the first tool specifically invented for bomb disposal. Inspired by a discharging cap that been dropped with one of the UXBs in Scotland, they realised that a live fuze could be drained of its charge if both plungers were depressed to make contact with the condensers and then earthed. To this end an apparatus was designed by the Research and Experimental Branch of the MOHS, consisting of a two-pin clamp that fitted on to the fuze head, depressed the chargers and discharged the energy within. The pins were worked by hydraulic pressure through an oil-filled tube 150 yards long – allowing it to be operated from a safe distance – and depressed by a grease-gun grip at the far end. It was the first of many such triumphs for British engineering ingenuity and scientific prowess.

A few weeks after Merriman had made his report the chance came to see if his theories worked for real. On 11 February 1940 a North Sea grain ship, *Kildare*, limped into Immingham Dock on the Humber. Her cargo had been displaced when a 250-kilogram bomb had penetrated the hold and gone off. Its twin was wedged in the after well deck cabin.

The call to RAF Manby was taken by John Dowland, a twenty-five-year-old flight lieutenant, but as he was leaving for Hull he was intercepted by the chief instructor and told to take Len Harrison with him. When they got aboard *Kildare* they found that the crew had bravely put mattresses under the bomb to stop it rolling about, but they were surprised to see that it had not one but two fuzes. Fortunately both were easily accessible.

This was to be the first time that anyone had tackled a UXB with a fuze thought still to be live. The Shetlands bombs had not been examined by Merriman until several weeks after they had fallen, by which time their charge had dissipated. It is unclear if Lowe had told Harrison about Merriman's conclusions, but it is apparent from Harrison's recollections that he used the same discharging technique as that devised by Merriman. It was also the one that he had recommended to Lowe back in the summer.

The forward of the two fuzes in the UXB seemed not to have been charged, so Harrison turned his attention to the one nearer the tail. He was aware that at any moment a movement of the ship might jolt the bomb against the cabin, exploding it and killing both him and Dowland. He therefore proceeded methodically but swiftly. Disarming the device 'necessitated my depressing the plungers, keeping one side of the fuze earthed while each plunger was checked. For this purpose I used an ordinary +/− 30 V voltmeter.' He next gave the voltmeter to Dowland, so he could confirm that the forward fuze was uncharged, and then unscrewed the locking rings with pliers. After the fuzes were withdrawn Harrison made sure to remove the gaines, as if even these were to go off they could kill by fragmentation.

A few weeks later, on Good Friday, Harrison was again called into action by the Humber authorities. A fishing vessel had come into harbour with a UXB aboard, so Harrison travelled up to Grimsby and had himself rowed out to the boat by two airmen. The missile was unlike any he had seen before. About a foot wide and shaped like a dustbin, it had jammed against the coming on the port side of the foc'sle. The fuze was clearly visible within a centrally-placed locking ring.

'Knowing the German words for "safe" and "fire",' he afterwards recalled, 'it appeared to me at the time that it would be relatively safe procedure to depress the fuze plate and turn the dial until the pointer lay against the word *Blind* on the side of the fuze perimeter.' Harrison accordingly made about seven or eight attempts to turn the dial, but it would not budge. The next step, he decided, was to break open the casing of the bomb rather than meddle with an unknown fuze. Inside he saw a greenish powder atop what looked like large cheeses with a hole in the middle through which the fuze passed. These he threw into the sea.

Having cleared out these explosives, Harrison could now get at the rear end of the fuze pocket. He tugged off the plug, pulled out the detonators and then, proceeding up the tube, removed the fuze itself. After signalling the airmen on the dockside that he was ready to return, he went to wash his hands in the cabin, only to find that his hands had turned bright pink. In his haste to get to Grimsby he had forgotten to bring the rubber gloves he usually wore, and had had to borrow a pair of woollen ones. He assumed that his skin must have had a reaction to the explosive powder.

Meanwhile, the airmen had not been wasting their time. They had managed to stuff into their kitbags a fair supply of the white fish that had been in the hold and which they thought might nicely supplement their rations at Manby. On reaching the quay Harrison was met by the beaming owner of the vessel, and with some embarrassment had to make the airmen confess to their pilfering. Fortunately the man was only too happy to make them a present of their catch.

The next morning Harrison was woken early at home by a naval officer who had been sent by the Admiralty to collect the fuze. He was loath to part with it as he had not had a chance to examine it properly, and was able to persuade the officer to let him photograph it before handing it over. Such understanding and cooperation, occasionally tempered by friendly rivalry, was soon to become the norm between the Services in the field, but antagonism higher up the chain of command would in time threaten the very success of their bomb disposal operations.

'As soon as you introduce a new weapon in war,' reflected Herbert Rühlemann, 'the enemy always gets ahold of it.' The secret of the electric fuze was now out and Germany's opponents were impressed. The British had not been the only ones studying it. The French had also recovered some examples (probably from Poland), and their report, found by the Germans when they took Paris, concluded that it must be very expensive to manufacture. The twenty or so joins between the wires in the fuze had been welded, when it would have been cheaper – but less reliable – to solder them. The French also assumed that it must require at least a thousand volts to fire the igniter.

In fact, the genius of Rühlemann lay in having created a fuze that was not only far more sophisticated and robust than any before, but that also cost very little to make, from the gossamer alloy for the igniter to the hand-held arc welder Siemens developed for the production line.

For his part, Rühlemann was glad to get some feedback about the effectiveness of his invention. He had only received periodic reports from Spain, but with the invasion of Poland, and those to come in the spring of 1940, he began to receive more information. From his account of his journey to Poland shortly after its surrender it is evident that he was proud of his achievement, if a little awed by the use to which it was put.

In one town, the Luftwaffe had carved a path of destruction 'several hundred feet wide and a mile long right through the middle . . . Hundreds of people were sifting the rubble in the hope to find some of their belongings. This was the first time I saw the results of the bomb raid, a small one.

'As the war advanced, I did see many, many more, and heavier ones.'

3

WARS ARE WON IN WHITEHALL

On a warm evening in June 1940 some of the most promising talents in British science were to be found clustered around a restaurant table in Soho. They belonged to a dining society that met regularly to promote better public understanding of science. Tonight, the talk was of the likelihood of an imminent German invasion, and why it was that so few scientists had yet been diverted into war work. Opinion was divided as to whether this was the fault of the Establishment, which regarded them as troublemakers, or that of the leaders of the profession, who were content to leave the conduct of the conflict to civil servants and politicians. All were agreed that more needed to be done to increase public awareness of the vital contribution that science could make to the war effort.

One of the scientists' guests, the art historian Sir Kenneth Clark, felt bound to defend the Government's use of the profession given that he was now at the Ministry of Information. So stimulating did the debate become that another of those present, Allen Lane, the founder of Penguin Books, said that it was a shame that no record had been taken of it. Straightaway the offer was made to deliver to him, within a fortnight, a manuscript based on the arguments advanced. Within a month of the meeting, this was published as *Science in War*. It was to prove both a best-seller and highly influential.

The guiding spirit of what was in effect the younger scientific community's manifesto for their greater involvement in a war that they felt was being lost for want of it was J. D. Bernal. At thirty-nine, he was still seen as the *enfant terrible* of British physics, known as much for his outspoken support of Communism as he was for his breakthroughs in X-ray crystallography that would pave the way for those in genetics in the fifties. His Left-leaning sympathies were shared by many of the other members of the dining club that had hosted the Soho discussion.

The Tots and Quots had been founded by Bernal in 1930, taking its name from a Latin remark meaning 'So many men, so many opinions' made by the biologist J. B. S. (Jack) Haldane at its first supper. However, most of those who attended later meetings shared a single conviction, namely that despite the remarkable advances made by scientists in recent decades they were still of little account in government thinking. State-funded research was largely reserved for a small number of prestige projects and few scientists were employed at senior grades in the civil service.

This conservative cast of mind appeared most prevalent among high-placed men in uniform. They seemed to trust their gut instincts over the findings of scientific method. Bernal's politics perhaps prejudiced him against them, but his pessimism about the role that science would be given in this high-tech war was predicated on experience. From the Munich Agreement onwards he had criticised the government's lack of readiness for the coming clash with Germany, and in the spring of 1939 he had been invited by Stradling to join the MOHS's Civil Defence Research Committee. For them, he had undertaken important experiments on the effects of explosions on shelters, but the reception given some of his suggestions had led him to conclude that 'In the general view, scientists who want to get things done are regarded as meddlesome and troublesome.'

That attitude would persist for some time, but in the months before the Tots dinner important innovations had begun to make themselves felt at the heart of the military machine. The focus of these changes, which were to have momentous consequences for bomb disposal, was like Rühlemann a modest, meticulous, world-class engineer named Herbert.

The world from which Herbert Gough came was not that of the officer class. The son of a lowly civil servant in the Post Office, he was born in 1890 in Bermondsey, then most associated with the stinks of the leather trade. His was a world bounded by two of the tenets of middle-class Victorian life: knowing one's place, and self-improvement through hard work.

By the age of thirteen he had already begun a vocational training for a post in the engine room of British industry. Lessons in carpentry, pattern making and machine tools at Regent Street Polytechnic Technical School made him familiar with the workbench, but Herbert also had a singular intelligence that, combined with determination, brought him a scholarship at seventeen to University College School in Hampstead. A year in more genteel surroundings may have added a little polish, but by nineteen he had fetched up as intended in the factory yard, as an apprentice draughtsman at Vickers, the armaments firm.

First at Barrow-in-Furness and then at Erith, Gough learned to draw, design and construct gun mountings and fire control gear, but all the while he was studying at evenings and weekends for a degree in engineering. In 1914 this freed him from the drawing office at Vickers, taking him instead to Teddington, and the Engineering Department of the National Physical Laboratory (NPL), the country's leading research institute for the practical application of physics. Within a year, however, he found himself in Flanders.

The war was to have a formative effect on both his character and career. He spent all of it in France and Belgium with the Royal Engineers. He was commissioned in the field in 1916 and as a lieutenant commanded a signal section until he was demobbed in 1919. During that period he was in charge of communications in various parts of the line, including at the Somme, ending his time in the Army at the Cologne bridgehead. He was twice mentioned in dispatches and appointed an MBE.

Gough returned to Teddington with his boyish looks unchanged but with knowledge of how the military worked, with experience of having led a team of men, and with renewed confidence in his own judgement. He was also boosted by his recent marriage. Over the next twenty years he would rise to the top of his profession.

The science that Gough would make his own was that of metal fatigue. This was a common problem in industry, which had need of practical solutions to the problem of failures in chains and cables. Gough was to make numerous contributions to the understanding of stresses in hooks, rings and other lifting gear, as well as to the causes of fretting and corrosion. His particular interest was in fatigue failure: that due to repeated applications of loads much lower than needed to induce immediate failure.

By 1933 Dr Gough, as he had become, was recognised as a world authority on the subject, and that year he was elected a Fellow of the Royal Society. He was now also Superintendent of his department, directly responsible to the Director of the NPL for its eighty engineering staff and all the projects they carried out in their laboratories and workshops. He was regarded as 'a conscientious worker with a high sense of responsibility' whose 'keen and critical mind made him outstanding in research and administration'. It was this combination of qualities that was to bring Gough an unexpected job offer, and a marked change of direction.

The Civil Service is not perhaps so intrepid a force as the Commandos, but it is a truth insufficiently acknowledged by military historians that wars are won not on the battlefield but in Whitehall. If there are too few helicopters to ferry troops over country in Afghanistan laden with IEDs, that is the consequence of decisions made, often years earlier, in London. The sinews of war – funding, but also foresight and planning – are what allow a nation to punch at, or above, its weight.

One of the lessons that had been absorbed from the First World War was of the need for the Armed Forces to keep up to speed with developments in technology. If a potential enemy built up too great a lead in any arms race it could very quickly prove fatal. There was, of course, much work being done in state establishments, as well as by private companies, but the perceived need was for this to be better coordinated. As early as the mid-twenties, it was recommended that the service ministries each appoint a dedicated Director of Scientific Research (DSR), but although the Admiralty and Air Ministry did so, it was not until 1938 that the War Office saw the need to follow suit.

The post was to be one with wide responsibilities, which would give the Army its own voice in weapons development. The DSR was to collaborate with and supervise all research at the War Office's technical establishments, notably Woolwich Arsenal, to represent the ministry on inter-departmental committees and to keep a watch on work abroad. The task of recruiting the right man was entrusted to the head of the powerful Directorate of Munitions Production, Engineer Vice-Admiral Sir Harold Brown, and to Frank Ewart Smith, the chief engineer at ICI, which supplied much of the Army's explosives.

It is a measure of how scientists thought they were likely to fare in government employ that of the eight names suggested by Smith only two would consider taking the appointment. Of these, Herbert Gough was considered the more suitable, although he had not been Smith's first choice.

Gough, of course, already worked for the government, and for him this represented a major step up the ladder. That was not, however, his main motive for accepting, as he made clear in a letter to Smith: 'First, service in such a post offers great opportunity for doing something really worthwhile for one's country. Secondly, the work affords such fascinating and intimate contacts with a range of problems which must extend into so many branches of scientific research and practical development.'

These do not appear to be merely customary sentiments. He had a sense of duty and, as his son wrote later, 'his work was his greatest joy'. The opportunity to marry the two attributes was irresistible to him. His correspondence also reveals Gough to be diligent and loyal, with an attractive sense of humour, but he was encumbered by a slight social inferiority. He might cultivate a taste for Gilbert and Sullivan – he sang baritone in NPL productions – and work at his golf and tennis, but even these enthusiasms would be regarded by the grander officers with whom he would have to deal as irredeemably middle-class, the mark of the jumped-up clerk. With a background in science rather than in the classics or in the saddle, Gough in government was an outsider.

The theme of much of his time as DSR was to be that of overcoming an obstinate rearguard action by the Army's own specialists to any scrutiny by a civilian. It was not that the Armed Forces did not appreciate the possibilities offered by science. For example, in the early days of the war they swiftly grasped the advantages conferred by radar. However, the officers responsible for research and development believed that they, as professional soldiers, had a better understanding of the Army's requirements than any layman.

The truth was that although these officers were experienced in the problems faced by troops they were not – unlike their air and naval counterparts – trained scientists. In the verdict of the official historian of wartime production, J. D. Scott, they were 'at the best gifted amateurs, and at the worst not gifted'. There was a fair-sized corps of technical experts in the research sections who were capable of important innovations, but they were not encouraged to stray outside the tasks defined for them by their military supervisors.

On the modern battlefield, such narrow vision was to prove an increasingly dangerous weakness. For instance, tank designs in the thirties were, in the interests of spreading work around, commissioned from different firms each year. Some built prototypes with effective armour, others those with powerful guns. A lack of money and expertise meant that none developed a tank with both, a failing that was quickly, and bloodily, exposed in encounters with German Panzers in North Africa. The production of weaponry was to be radically improved only when Frank Ewart Smith moved from ICI in 1942 to become Chief Engineer Armaments Design, leading to the promotion of research staff on merit rather than by seniority.

Accordingly, when Gough took up his new post in June 1938 he soon learned that those who carried out the research he was supposed to control were responsible not to him but to two brigadiers, the Directors of Artillery and of Mechanisation. They had little to do with each other and wanted even less to do with him. 'This, at any rate, was the picture as it presented itself to the new director,' writes the official historian, 'and he deplored it freely.'

The following year, the situation was further complicated by a significant change in the provision of munitions. A review of the production capacity of the service ministries concluded that (aside from the Admiralty) it fell below that which would be needed in wartime. As a result, the Cabinet approved in April 1939 the creation of a Ministry of Supply (MOS). This move had the backing of the War Office, which felt that the existence of the new ministry would free it to get on with its proper function of soldiering, but it was resented by the Admiralty and Air Ministry, which wanted to retain control of their construction and design needs. In the event, the MOS would prove vital to the success of the hugely enlarged wartime forces, arguably playing a larger role in the conflict than the service ministries themselves. Nevertheless, the wholesale transfer to the MOS of much of the existing supply organisation on the Army side, including Gough and his military colleagues, on 1 August 1939 could only increase the intransigency of the latter. They were now beholden still further to civilians.

Matters came to a head in May 1940. In a letter to the professional head of the ministry, its Permanent Secretary, Gough asked

for 'a clear statement to be issued defining the duties and responsibilities of this Directorate'. His concern was that the setting up
of the MOS and its new departments had weakened any hold he
had managed to get over experimental work, making it easier for
the military to sideline him. He was, for instance, keen that the
tests being conducted by John Cockroft, the future pioneer of
nuclear power, at the Radar Research Station in Malvern be
supervised by him and not by the Army.

Gough's complaint was symptomatic of wider tensions within
the MOS, which had yet to bed down properly. They were
acknowledged a few weeks later by the new Minister of Supply,
Herbert Morrison, who told the House of Commons that the
brief time he had spent in the job so far had been the 'most
strenuous and complicated' of his life. Like Morrison, who was to
make wide changes in an effort to lessen the deadening grip of the
Army, Gough was a pragmatist, more interested in finding solutions to urgent problems than in the issues of status and
demarcation that troubled career officers.

'My outstanding impression of him,' recalled a colleague from
this time, 'are his sense of order, his vigour and resilience, coupled
with combativeness, forcefulness and drive; qualities which were
greatly needed in those years and he gave them in good measure.'
Nothing exemplified this better than Gough's involvement with
bomb disposal, which was about to let him show the Forces
exactly what civilian scientists could do.

In early February a decision had been made that the Home Office
was not capable of creating the disciplined force needed to deal
with any UXBs that might fall. Instead, they were to continue to
be the responsibility of the War Office. A consequence of this was
that the Ministry of Home Security, which was also run by the
Home Secretary Sir John Anderson, approached the MOS to see
if – with its connections to the Army – it might take on the
MOHS's bomb disposal functions. Initially the intent seems to
have been to confine these to the manufacture of equipment, the

MOHS not being a production department, but by 1 April Gough had agreed to oversee all 'necessary further research and technical development arising in connexion with the various methods for dealing with unexploded bombs'.

Although the conventional view is that Gough and his team began their war against the bombs from scratch, in fact many of what would prove to be their key tactics had already been devised by Reginald Stradling. In a series of meetings and reports to Gough in the spring of 1940, he summarised the situation so far.

A lack of examples of UXBs – though in some ways welcome – was, he said, still hampering better understanding of them. (The first civilian was not killed by bombing until 16 March 1940, and that was in the Orkneys.) Nonetheless, it was anticipated that they might become progressively more sophisticated and dangerous: 'No details of any German long-delay fuzes are known, neither is it definitely established whether in fact they are available, but the possibility of their use cannot be ruled out.' If they were to be found in future, it was thought that they were 'most likely to be used with large bombs and to be associated with a booby trap'.

As Stradling pointed out, 'the handling of an unexploded bomb is not difficult in the absence of a long-delay action fuze', as Merriman and Harrison had shown by short-circuiting the condensers. Nonetheless, since it was impossible to gauge what type of fuze was being dealt with, it would be safer to treat them all as if they were long-delay and protected by anti-interference devices. So while work was still going ahead on the discharger, Stradling had also begun to look for a way of making the bomb safe without having to touch its fuze. The obvious answer was a technique that removed the explosive.

Various methods had already been tried, some of which were more promising than others. The first step was to open up the bomb, and to do this one had been placed 'in a bath of liquid air with the object of rendering the case sufficiently brittle to break up from a blow with a hammer'. When it was taken out, the bomb needed several hefty blows from a sledge hammer before the steel fractured. Given the potential presence of a booby trapped fuze, this was thought 'not encouraging'.

More hopeful were the trials with chemicals undertaken by Professor Ulick Evans of Cambridge University, an expert on corrosion, who was using acid assisted by an electrical current to make a ring-shaped cut through the bomb. Attention had also been given to means of detonating a UXB in such a way that the explosion would be far less intense. The most successful of these involved thermite, a metal oxide used to generate high temperatures in welding. The theory was that, having lanced through a small spot on the casing, the thermite would burn off much of the bomb's filling before any explosion.

This idea had come from the Civil Defence Research Committee, of which Desmond Bernal was a member, and it was he who claimed the credit for having suggested another approach to Stradling. 'I can still remember,' he wrote afterwards, 'during a discussion early in the war on coping with the fuzes in unexploded bombs, that I hit in a few minutes on a solution derived from schoolboy egg-blowing.

'It consisted simply of emptying the explosives out of the bomb with steam from a safe distance and thus making it harmless regardless of the nature of the fuze.' Though he may have been the first to propose this plan, it was undoubtedly refined by others, most notably Gough's successor as engineering superintendent at the NPL, S. L. Smith. By the autumn of 1939 he was already investigating this 'most promising method of attack' and by the time of the handover to the MOS he had built a working prototype.

This was a 'mechanical device . . . a trepanning tool driven by an electric motor or steam engine and mounted directly on the bomb. After one or more holes have been cut in the casing the drills are automatically removed and a steam jet is fed into the casing to melt and remove the explosive.' Heartening though the development of this contraption appeared to be, some important questions about its use remained. These included the small matter of the effect of the heat and pressure of the steam on the bomb's charge and detonator.

In passing on the baton, Stradling and the MOHS had certainly given Gough a healthy lead. In measuring the achievements of the next few years, however, it needs to be remembered that as yet no equipment for bomb disposal had been manufactured, much less used on a live bomb. The MOS's entire experience of UXBs, meanwhile, consisted of a single disassembled electric fuze in the Research Department at Woolwich. There was much to be done.

The challenge was about to get far stiffer. By 1940 Rheinmetall's new plant at Breslau was coming on stream, with Rühlemann installed as its technical director. With the help of more than a thousand 'guest workers' supplied from France, Czechoslovakia and the Balkans by the SS, production at the factory was ramping up. At its peak, output would approach three million fuzes per year.

Rühlemann later declared that he tried to do his best for the *Gastarbeiter*, but was afraid of the SS. Even so, when writing his memoirs forty years after the war he made no condemnation of the Nazis beyond that they were rather uncouth, a shortcoming he attributed to a lack of education. Like many who worked for them, he seems to have closed his eyes to the horror and focused on the job in hand. He was proud of his team – Berger at Mechanical Fuzes, Übelmann at Electrical, Riedel in charge of R&D. 'We were a group of dedicated people working for a common goal,' he wrote, as if theirs was just any other business.

Although stated in perhaps too generalised terms, the thrust of Bernal's criticisms at the Tots dinner was correct: the machinery of government was out of touch with advances in science. One proof of this was one of the few changes that Gough was able to make in his year at the War Office: the establishment of an advisory council of scientists reporting to the Secretary of State. Even so, this did not start work until 1940, by which time it came under the auspices of the MOS.

There, Gough was to be central to galvanising science for war. No sooner had he moved into the MOS's hastily requisitioned headquarters at the Adelphi on Savoy Hill and in the adjoining Shell-Mex building on the Strand than he began placing research projects with the universities. He also drew some of the leading younger authorities on physics and chemistry to the Ministry as in-house experts.

For a time these included Patrick Blackett, a Tot and a future Nobel laureate. In his early forties, Blackett would work for the MOS on improving the effectiveness of anti-aircraft fire before moving to Coastal Command. There, his use of mathematical models and scientific theories would vastly improve its successes against U-boats and the Luftwaffe, validating the use in war of what became known as operational research.

Another two Tots recruited by Gough were Jack Haldane and Lawrence Bragg. It was the former whose exaggerated accounts of the bombing in Spain – influenced by his Marxist beliefs – had largely led to the distorted predictions of civilian casualties in Britain. A renowned biologist, Haldane was to test early frogman apparatus to ascertain its effects on the lungs and the blood. Bragg meanwhile had been, at twenty-five, the youngest winner of a Nobel Prize, having collaborated with his father in developing the X-ray spectrometer that revealed the arrangement of atoms in crystals. (He would later play an important role in the discovery of the structure of DNA.) During the First World War, Bragg had developed equipment that measured the speed of sound waves and was used to reveal enemy artillery positions.

More closely involved in bomb disposal problems were two other eminent names. Ian Heilbron was a product of Jewish ancestry and a Scottish education. By 1939, aged fifty-three, he was Professor of Organic Chemistry at Imperial College, London, and known for having done much of the early research on vitamin A and on penicillin. He had military experience too, having won a DSO in the First World War. He was to make a significant contribution to the conduct of the Second by championing the pesticide DDT as the solution to the losses of manpower from lice and mosquitoes in Africa and the Far East.

A meticulous and energetic man, with little patience for delay or for slack thinking, Heilbron was often able to see with exceptional clarity an answer that had eluded others. He also had a gift for predicting the reactions of colleagues, which made him a powerful presence on committees. Almost all the way through the war he slept on a bed made up in his laboratory at Imperial, allowing him to continue running his university department as well as helping the MOS.

Gough's principal adviser was a physicist, Edward Andrade, known universally as Percy. Like Heilbron, he was both Jewish (though a Londoner) and had studied in Germany before the Great War. During the war, he had been twice wounded at Arras while working on sound-ranging and counter-batteries with Bragg, whom he knew through his research in Ernest Rutherford's laboratory on the wavelengths of X-rays. His main interest, however, which was a point of contact with Gough, was in the mechanical properties of matter, especially the creep of metals.

Again like Gough, he was now in his early fifties, and was Professor of Physics at University College, London, having previously taught at the Ordnance College, Woolwich (now the Military College of Science). Yet work was for him not all that life offered. A champion boxer in his youth, and a keen cricketer and walker, he also had a passion for literature, and knew Hilaire Belloc, T. S. Eliot and Somerset Maugham. He was clubbable, liking nothing better than long Saturday lunches at the Athenaeum. And he was an expert on, among other subjects, witchcraft.

Andrade's chief talent was the ability to explain the complex in simple terms, which from 1942 made him a popular guest on the BBC's *The Brains Trust*. He also had a sense of fun: each Christmas he gave a lecture at UCL in which the experiments were set up to confound his pronouncements. At the MOS, he was to have a key role in scrutinising the many inventions and suggestions submitted by members of the public.

It may have been difficult for Gough to marshal these large reputations and egos, but the skill and determination with which he did so was admirable. Nothing better illustrated his capacity to elicit fruitful collaboration than his management of one of the war's most successful uses of intellectual firepower.

The Unexploded Bomb Committee should, by rights, be almost as well-known as the Twenty (Double-Cross) Committee – the team of scholars and intelligence officers who turned German spies – or the collected geniuses of Bletchley Park. Its work was no less secret than theirs, and at times no less vital. It may have since been neglected because its members never wrote about it, or perhaps because it lacked the frisson of espionage. Yet it far exceeded its more celebrated contemporaries in terms of dangers run, courage demanded and sacrifices made. The Committee and the small band of civilian staff that served it were between them to win three George Crosses, a George Medal and two Commendations for Bravery. Four of those awards were posthumous.

The idea to have a committee 'to co-ordinate experimentation' on fuzes was originally Stradling's. It was gladly adopted by Gough as a sensible way of pooling knowledge from all the branches concerned with bombs, and of keeping all abreast of developments, and the committee's terms of reference were broad: 'to consider the general problem of the unexploded enemy bomb and to advise on methods of dealing with the problem'. Gough sent out invitations to join the committee ten days after taking over from Stradling, and three weeks later, on 1 May 1940, it met for the first time at Savoy Hill House.

Gough, who was to chair the committee for the whole of the war, was the first to speak to the eighteen people grouped around the table. Having assured himself that they had all signed the Official Secrets Act, he rapidly outlined what was known of German fuzes. Then he introduced its joint secretaries, Merriman – whom he had poached from Stradling – and an Army captain.

There followed a general discussion about some of the issues that had been thrown up by the MOHS's work. Stradling spoke, as did Dr Robert Ferguson, the representative of the Research Department at Woolwich, and one of Gough's nominees, Professor William Garner. Although he now taught at Bristol University, he had spent much of his career at Woolwich doing the experiments that had established the fundamental facts underlying the chemistry of explosions. Self-effacing and devoid of ambition, Garner was happiest tending his orchids.

The committee agreed that if time was of no importance it would have concentrated its efforts on developing a machine which would remove the explosive from UXBs by remote control. Yet, since time might prove critical, it was decided to proceed instead on three fronts.

The first of these was the fuze discharger, a pocket version of which the MOHS had already been developing. It took the form of a shallow-rimmed brass disc about one and a quarter inches across, from which projected two short parallel prongs. These depressed the plungers, the device being held in place on the fuze by a sidescrew. Two weeks after the Unexploded Bomb Committee (UXBC) met, the fuze discharger became the first piece of bomb disposal equipment approved for use. Two and a half thousand of them were ordered from J. A. Crabtree & Co. of New Oxford Street, and so they became known as Crabtree dischargers. The cost of each was three shillings; the number of lives and the value of property saved by them incalculable.

Characteristically, Gough had pushed through its manufacture by simplifying the MOHS's design. What he sought was not the perfect solution to a problem, but something that would do the job immediately. In addition to the discharger, he ordered another tool that would become central to the bomb disposer's armoury, an adjustable key with which to unscrew the fuze's locking ring.

Just before lunch, having seen a model of Smith's steam blower (which, it was agreed, should be the second strategy pursued), the

committee was treated to a surprise. Moxey, Merriman's com-
panion on the Orkney adventure, was shown in and assembled a
peculiar-looking contrivance on the table. It was all screws and
pistons, and its purpose was not immediately apparent. And was
that a Sparklet bottle, as used to put fizz into drinks?

Moxey then demonstrated what he called his automatic fuze
remover. Even if a fuze had been discharged there might be cases
when it was thought that behind it was concealed a trap. His
device would negate the need to withdraw such a fuze by hand.
The mock-up consisted of a saddle that could be clamped directly
on to the bomb atop the fuze. Above the saddle were two cylinders that contained a piston which could be attached to the fuze.
This was pneumatically driven by the Sparklet bottle: as the compressed air passed into the lower cylinder the piston slowly rose
and the fuze was pulled out without anyone needing to be nearby.

The UXBC liked what it saw, and after a few modifications to
make it simpler and stronger Gough ordered 250 fuze extractors –
and a thousand bulbs from Sparklet Ltd of Edmonton – at the end
of May.

The last significant contribution at the meeting was made by
Andrade, whom Gough had favoured as a member over
Stradling's suggestion of Bernal. He and Merriman had been
thinking of ways of neutralising any explosive in the fuze pocket
so that a booby trap would not function. He suggested that it
might be worth trying to do this by pumping some form of
'desensitising liquid' into the pocket. It was agreed that this technique was worth exploring further. At first it was to prove
something of a false hope, but much later in the war the method
would prove to be the answer to almost every bomb disposal
conundrum, and is still the basis for much modern BD work.

Within a few weeks of taking up the reins, Gough had started
to foster a strong spirit of cooperation between scientists, design-
ers, manufacturers and end-users of the equipment they devised.
This ethos would become the hallmark of the administration of
bomb disposal, and an example to the rest of the Civil Service.

The ground had rapidly been surveyed, plans drawn up, materials ordered and problems anticipated. It seemed as if the future could be faced with some confidence. Now it was up to others to do their bit.

4

I MAY NOT LAST VERY LONG AT THIS

While the scientists set to work in the early months of 1940, progress was also slowly being made in deciding how the threat of UXBs would be dealt with on the ground. At the War Office, the file had now landed on the desk of Major-General Ken Loch, Director of Anti-Aircraft and Coast Defence. In mid-March he submitted a report proposing the creation of a specially trained force of independent squads. 'A convenient title for this basic unit,' he suggested, 'is a Bomb Disposal Section.'

These were intended to replace the three-man contingents who had been marooned since their hasty deployment in the winter. Loch thought that 109 sections, each consisting of an officer and sixteen other ranks, should be sufficient for requirements. They would be divided into two sub-units, one to undertake removal – 'i.e. a digging party' – and the other to render the bomb safe. They would be equipped with basic tools such as rope, picks and crowbars, and with transport, namely a motorcycle and two lorries.

Loch's idea was that the sections could be formed largely from retired soldiers, and that this would be a full-time organisation but not part of the regular Army. Minds were then concentrated by the first bomb to fall on the mainland, near Canterbury on 9 May, and then two weeks later by the first raid on a large town, Middlesbrough. By late May it was accepted that the work would not suit pensioners, and the commanding officers of the School of Military Engineering at Chatham and of the four Royal Engineers Training Battalions at Shorncliffe, Newark, Chester and Colchester were told to raise an initial twenty-five sections.

Nonetheless, they were told that 'personnel for these sections should be selected as far as possible from men of low medical category including those fit for Home Defence only'. The nature of the task ahead had yet to be appreciated, and with men needed in France this was not considered the equivalent of a front-line posting. A list of sixteen officers was drawn up to take charge of the first sections that paraded in June. Of the names on it, half would win the George Cross or George Medal; a quarter would be dead before the year was out.

The establishment of this force had led to a better demarcation of bomb disposal responsibilities between the three Services. The Royal Navy, which had already suffered some casualties in attempts to recover and neutralise shipping mines, would deal with any bombs that landed aboard ships or on Admiralty property, such as dockyards. The RAF would look after UXBs on airfields and in aircraft that had crashed. The Army would deal with everything else.

With their experience of their own bombs, the Air Force was naturally seen as the leader in BD procedure. A number of so-called X Stations had been set up at key aerodromes, commanded by senior NCOs who would organise the removal of any threat. Len Harrison had already begun to train them at Manby, but information about UXBs was still sparse. 'They were given what we had on German bombs and fuzes,' he recalled, 'the accepted methods of dealing with them, and of implementing safety measures on the station.'

From May onwards, with the Germans now in Holland and Belgium and the BD structure being expanded, Royal Navy and

Royal Engineers officers were accepted for training in small groups at Manby too. The course was to be held there until August, when due to the increased intake at the air gunnery school it was moved to RAF Melksham in Wiltshire.

The syllabus had been drawn up at the end of 1939 and took account of all the information passed to the Air Force by the Armaments Department at Woolwich and by HMS *Vernon*, the Navy's torpedo and mining school at Portsmouth. Even so, the course barely stretched to a week, with just ten hours devoted to the construction and defuzing of German bombs. It had been hoped that more material would come out of France, but the French appeared not to have grasped the workings of fuzes. Hapless navvies had dug for buried bombs, with fatal consequences. It was anticipated by the British that most UXBs could be demolished in situ, so much of the course dealt with locating and avoiding damage to underground service pipes and mains.

One of the early trainees was William Wells, who had been given an emergency commission in the RE and sent to Manby in June. He was impressed by the modernity and whiff of glamour that seemed to emanate from the young and keen RAF. 'We found the aerodrome a superb place,' he wrote. 'It was new, the buildings and quarters very attractive in red brick and stone, in contrast to the dingy group of buildings which comprised Kitchener Barracks at Chatham.' While he was there, he heard the news of the surrender of France. 'Nobody seemed to worry very much: both the RAF officers and our party accepted the situation with nonchalance.'

On the course with him was an Irishman named Max Blaney, whose exuberant manner and jokes endeared him to Wells. Afterwards, they travelled down to London, where Wells found that he was to be posted to Glasgow. He travelled north with little more than his rudimentary BD tool kit of a watchmaker's screwdriver, wire cutters, pliers, spanner, Crabtree discharger and a tin of Vaseline.

The scanty nature of his training contrasted sharply with that long enjoyed by the Germans. Their bombs were the concern solely of the Luftwaffe, and until 1939 its officers and NCOs alike qualified as armourers only after passing a two year, highly technical course held at Halle, near Leipzig. Following the out-break of war, this was compressed in 1940 into seven months. It was graduates from Halle who were later to deal with all the unexploded bombs that landed on Germany, liaising with a civilian organisation made up principally of former armourers or *Feuerwerker*. When the bombing came, it was undoubtedly the Germans who were at first far better placed to counter the dangers of UXBs.

One of the curiosities about the first batches of officers selected for BD was that many were middle-aged with a professional qualification. Most were newly commissioned rather than regular soldiers, although some of the older men had fought in the Great War. That may explain why they were selected for BD duty by the RE, which did not want to draw on its core fighting strength, but not why their areas of expertise usually had little to do with explosives. In the way of the Army, any faintly relevant experience seemed sufficient to recommend one for Bomb Disposal. The joke that soon went round, once it was known how dangerous the work was, was that the generals preferred not to expend career officers expensively trained in regimental duties on it.

Among those who found himself in BD almost at random was an architect, Stuart Archer. The son of an electrical engineer, Bertram, and a demanding mother, Frances, he was born at Abbey Road, London, in February 1915. The household was comfortably off: his sister was able to train as a ballerina, and from their home in Hampstead he could walk to the classes and rugger pitches of a preparatory school education.

All this changed when his father lost much of his money in a bank failure. At thirteen, Stuart found himself instead, like Herbert Gough, at the Regent Street Polytechnic, earmarked for vocational training. What spared him the factory office was a talent for drawing, which had emerged in the form of illustrations copied from his children's books. Pushed on by his mother, and with his father making what sacrifices he could (including, at times, his sister's ballet lessons), at fourteen Stuart was articled to a firm of architects in Gray's Inn. The awareness, from such a young age, of his parents' expectations of him inevitably gave rise to a strong sense of duty.

Archer passed his exams as soon as he could, at twenty-one, and joined the practice of Ingram Son. It specialised, he would say, 'in licensed premises – in other words, pubs'. Most of the projects were for London breweries, notably Young & Co. Perhaps the best known of those he worked on was The Cricketers, on the green at Mitcham, and his prospects seemed good when, in 1938, he was made a partner in the firm.

Others, however, had also been making plans. 'One could appreciate that a war was going to come, and you wanted to support your own side.' So Archer did 'the right and decent thing', and that same year joined the Honourable Artillery Company (HAC), the Territorial unit based in Moorgate, on the edge of the City of London. Usually the TA was only too glad to have recruits, but Archer was conscious of being carefully scrutinised at his interview by the young officers on the other side of the table, most of whom had rather grander jobs than he. Archer went in as a private, and was soon being drilled once a week by instructors from the Grenadier Guards. 'They spoke very loudly and clearly, and we ended up by doing exactly what they required us to do, and we did it smartly. That was the whole object of it.'

His father had not fought in the First War, but two uncles had. Archer and his friends never felt that they lived in its shadow, but by the mid-thirties they were all talking about the next war. With the invasion of Poland on 1 September 1939, it arrived. A few hours later Archer was told to report to the HAC.

The long-awaited mobilisation proved something of an anti-climax. Archer arrived at the same time as a thousand other young men, and since there was nowhere for any of them to sleep they were all sent home. The following afternoon he made his way to

Bunhill Row again, and was told to come back on Tuesday. Everyone was waiting to see how Hitler would respond to Britain's ultimatum. Everyone except Stuart Archer, who was waiting to see how his fiancée Kit would take to the idea of bringing their wedding forward – to the next day.

'It seemed to me that now was the only opportunity to do it,' he recalled. 'So that evening I had to find out where I could get a special marriage licence. I went to the local town hall, but they had given so many out to people who had had the same idea that they had decided to stop. I was told that the only place that I would find one was at Westminster Abbey, but when I arrived they too were turning people away.

'But as luck would have it, I was in the HAC uniform and the chap there was ex-Army and had some sympathy for me. So I got the last licence issued, the last in peacetime.'

Kit was 'rather surprised' to be told what Stuart was planning, as was the vicar of his parish church, in St John's Wood, but both agreed to the sudden wedding. The final task that he had was to find a jeweller that was still open at eight in the evening, but fortunately many kept late hours on Saturdays. The ring cost thirty-eight shillings and sixpence.

The next morning he set off in his car to pick up Kit from her parents' house in Edmonton. Then, suddenly, he heard the sound of the air-raid alarm. Neville Chamberlain had told the people of Britain that they were at war with Germany, and no sooner had he finished speaking than the sirens began to moan. Wardens ran out into the street down which Archer was driving. 'They were trying to wave me to the side and saying: "Take cover, take cover!" and I was saying: "Get out of the bloody way, get out of the bloody way!" I went right through them.

'When I got to Kit's house they wouldn't let us out again, so I was trapped there until about half one or two. Then the air raid was cleared and that was that, so we drove over to St John's Wood and had the ceremony. So that in itself was quite an adventure, wasn't it?'

After two days with his bride Archer reported to his regiment. They marched out of the City and over London Bridge with bayonets fixed, and then took the train to Bulford Camp on Salisbury Plain. There, almost his whole battalion was enrolled on an infantry officer training course, although because of his architect's experience he was soon offered a commission in the RE instead.

Archer spent the next three months learning the art of the sapper: how to lay out trenches, how to build dug-outs, how to throw bridges across rivers, and how to blow them up. The length of the course had been halved because of the war, but by the end of it 'we had, we all thought, been well and properly trained'. Over-indulgence in drink, which Archer regarded as inappropriate, led to some of the cadets leaving early, but among the others he made strong friendships with two fellow transplants from the HAC. Fate was soon to sunder this bond: 'They didn't last terribly long, I'm afraid, because they were in fact both killed.'

Archer, meanwhile, was confronted with nothing worse than tedium. Expecting to be sent straight into action, in the spring of 1940 he was posted instead to Hemel Hempstead. There he was attached to a Territorial RE unit, 553 Field Company. 'They already had officers, so I was a bit of a spare person they didn't want.' He spent several frustrating weeks trying to span the Thames with a Bailey Bridge. Then, out of the blue, he found himself in Bomb Disposal.

'They were looking round for people to do it, and as a supernumerary in 553 Coy I was eminently suitable as they wanted me the hell out of there.' With a dozen others, he found himself in June at Manby. The course, he remembers, lasted four days, of which three were spent looking at British bombs. The high point came when their civilian instructors placed a fifty-kilogram German bomb on the table in front of them. It was about three and a half feet long, and nine inches around, with a pointed nose and steel fins at the back.

'This,' said the instructors, 'is a German bomb. It is the *only* one we have. And this is the fuze, the *only* one we have seen.' There was a quick demonstration of how to use the Crabtree discharger, and then the officers were passed fit for bomb disposal duties. As Archer swiftly concluded, 'At that time we knew sweet damn all about anything.' Looking back, he reflected that 'I think really people were thinking that the way to do bomb disposal was to get yourself a wheelbarrow, sweep them up and take them away.

'Nobody really knew what the bombs were and how they worked. When I started was when I learned the various improvements that the Germans had made time and time and time again.'

He was sent first to Chelmsford, where he gathered an RE serjeant, a lance-serjeant, two corporals and ten sappers, and then travelled on to Cardiff. Following their conquest of France, the Germans had begun to prepare for the invasion of Britain by significantly increasing the number and weight of their air attacks, launched from bases just across the Channel. Among the first targets were the oil refineries and ports of the South Wales coast.

With Archer went a new friend and fellow subaltern, twenty-year-old Edward Talbot. The youngest of six children, he came from a long line of well-born clergymen, the Chetwynd-Talbots. His father had had hopes of becoming Bishop of Shrewsbury, but had to content himself with the living of Egmond in Shropshire. After Harrow, Edward had only had a year at Cambridge, studying engineering, before war intervened. 'I got on extraordinarily well with him,' recalls Archer. 'He was a very nice chap, and knew as little as I did about the whole shooting match. He was willing to share whatever we were doing together.'

He and Archer were to command 104 and 105 BD Sections, both to be quartered at Fitzhammond Embankment, the Territorial Army building next to Cardiff's football ground. On their first night, however, they were to be billeted at the headquarters of the Welch Regiment in the centre of the city. Having gone to bed, Archer was shaken awake by what was to become the familiar shudder of bombs falling. So tired was he that he fell asleep again, only to be awakened by one of the Welch officers shaking his arm.

'You're Bomb Disposal, aren't you? Will you come down, we

think we've got a UXB at the docks.' Archer swiftly dressed and was driven the few hundred yards to the water. There, as advertised, was a hole in the side of the docks.

'It was about fifteen inches in diameter and went down about eight or ten feet. And I, thinking back to that one bomb lying on the table, said "Yes, I suppose that *is* a UXB", not having seen anything like it before.

'We went back to the HQ and woke up the serjeant and the men, got them dressed and started to march over there – we had no transport. And while we were doing that, the bomb went off. It was a delayed-action bomb, and went off before we got there. So that was Bomb Number One . . .

'I thought if it's going to be like this then I must try to get my wife to join me. We hadn't been married many months. I phoned her up and said that I may not last very long at this, so you'd better come down pretty quickly.'

Kit Archer put all her possessions in one half of the bassinet, the new baby in the other, and set off for Wales. Like the wives and sweethearts of all those in BD, life for her was about to become very uncertain indeed.

5

WILD JACK

As quietly as they could, the crew of the collier *Broompark* weighed anchor and eased off from the quay into the wide waters of the Gironde. Dawn was breaking over Bordeaux, and France was falling.

The tall, unshaven Englishman in the smart flannel trousers and dirty trench-coat had kept them hard at work all night, loading crates and helping several dozen frightened passengers come aboard. There had been pandemonium on the dockside the previous day. Crowds of refugees had scurried back and forth looking for a way of escape, then thrown themselves to the ground as the ship next to the *Broompark* had suddenly erupted in flames.

The crew had wanted to leave straight away, but the man had angrily brandished a riding crop, revealing brawny arms covered in tattoos, and then plied them with champagne until they were too drunk to care. Now he was dispensing more of it to the women on deck, telling them, 'This is the perfect remedy for seasickness.' From time to time a pretty blonde leaned over to light the cigarettes he smoked through a long black holder.

The more astute among the party might have discerned unusual grace and dignity beneath his high spirits. They may have noticed, too, the occasional glances at the tarpaulin that covered something lashed to a wooden pallet. The aristocratic demeanour could be explained by the revelation that their boyish host, now flourishing a pair of revolvers whom he called Oscar and Geneviève, was the Earl of Suffolk – known to friends as Wild Jack Howard.

Still more surprising was that hidden under the scruffy canvas tarp was a secret that would change the course of the war: the world's entire supply of heavy water, the key to the atomic bomb.

Wars create legends. Their intensity brings to the fore remarkable people, but they also encourage exaggeration and myth-making. Much that has been written about Jack Howard since 1940 has been unabashed speculation or distortion for the sake of a better story, but there has never been the need to indulge in either. Even his contemporaries were aware that he was simply extraordinary, blessed with brains, charm and courage – as well as individualism bordering on eccentricity – that brought to mind the heroes of the Spanish Main or the Scarlet Pimpernel. They were qualities that were to be spent in full in the service of bomb disposal.

As in all the best tales, Jack's past was both romantic and tragic. His parents had met in India. He was a dashing sportsman and ADC to the Viceroy, Lord Curzon. She was visiting her sister, who was Curzon's wife. Henry had a title and an ancestry that dated back to the routing of the Armada. Daisy possessed beauty, a love of adventure and one of the largest fortunes in America. They were married in 1904 and Charles Henry George – always known as Jack – was born to them two years later.

Their life together seemed to have all the ethereal glamour of Daisy's portrait by John Singer Sargent, an idyll of safaris and jaunts in the newly invented aeroplane, which she found thrilling. It all came to an abrupt end in 1917 when Henry Howard was killed in action near Baghdad during fighting against the Turks. Aged eleven, Jack succeeded his father as 20th Earl of Suffolk and 13th Earl of Berkshire.

More than ever now, Daisy was thrown back on the wealth of her father, Levi Leiter, co-founder of the Marshall Field department store and owner of some of the most lucrative property in Chicago. Unfortunately, although Levi was rich he was difficult, and though Daisy was rich she was exceptionally difficult. Growing up at the family seat, Charlton Park in Wiltshire, Jack was often caught in the whirlpool of his mother's moods. Her seeming resentment of him hardened still further when he injured his brother Cecil in a shooting accident, leaving him with only one foot.

Until then, Jack had largely been educated at home, notably by a French governess who had taught him to speak her language fluently. Now, however, he was packed off to Osborne, the naval college on the Isle of Wight. It was not a good choice of school for him. Jack was unhappy there, and began to show the first signs of what would become a disregard for convention and a dislike of excessive discipline. In 1921, aged fifteen, he was moved to Radley, which proved more sympathetic to his temperament. He started to demonstrate an unusual quickness of mind as well as a practical bent that found an outlet in the laboratory and the workshop. His most formative experience, however, came at seventeen when, on leaving school, he spent four months helping sail the clipper *Mount Stewart* from Liverpool to Sydney.

The time he spent in Australia helped him to find his feet by giving him a measure of independence, but Suffolk was nonetheless intended for the traditional life of an aristocrat. On returning home he joined the Scots Guards but, following a serious bout of rheumatic fever, was invalided out in 1926. He was often to be in pain from rheumatism, especially in his back, but would make light of it.

Suffolk was to spend much of the following decade back in the heat of Australia. The skipper of *Mount Stewart*, Captain McColm, had become a close friend and fatherly influence, and they went into partnership in a large sheep station in Queensland. Suffolk was a fine horseman and, according to a newspaper profile of the time, 'entered freely into the life of the land, and those sports that are part and parcel of life outback'. These did not include hunting or shooting, however. He abhorred violence, perhaps because of the accident to his brother, and hated causing any kind of pain to animals. His reluctance to do so, despite the comments that must have attracted in the tough world of the outback, was another indication of his growing self-confidence.

While his move abroad put some welcome distance between himself and his mother, she had her own reasons for being happy to see him go. Latterly he had been spending too much time hanging around stage doors for her liking, and seeing too much of Mimi Crawford – one of the best-known singers and dancers in the West End. Although her uncle was Lord Chalmers, a former Governor-General of Ceylon, Mimi's family were less respectable stage folk, her father having been manager of the Alhambra Theatre. And, despite her ingénue looks, she was eight years older than the smitten Suffolk.

Mimi had been at the height of her fame in the era of the flapper, whose styles suited her girlish figure. Having appeared in revues with the likes of Fred and Adele Astaire, Max Wall and Jesse Matthews, she was familiar to many both as a performer and as a cigarette-card pin-up. In 1929 she reached a still wider audience when she starred in a short film, *In An Old World Garden*, that was shown in cinemas in a double bill with the first full-length British talkie, Alfred Hitchcock's *Blackmail*.

Despite the distance, and his mother's opposition, Jack and Mimi remained in contact and in love. In 1934 he surprised everyone by returning to England, proposing to her, and a few weeks later getting married. The wedding was not held at the family church but in London.

Much to Daisy's displeasure, there were now two Countesses of Suffolk and two potential mistresses of Charlton Park. Daisy was determined not to be usurped by a younger woman, still less by one whom she did not consider a suitable wife for her son. She intended to bring him to heel, but was met by his firm resolve.

Tantrums ensued, and the Earl was forced to buy back contents from the house, which his mother was auctioning off on the lawn. Once again he was obliged to move away. This time he went to Edinburgh, having, at the age of twenty-eight, rediscovered his enthusiasm for science.

For the next three years he studied pharmacology at the university, graduating in 1937 with first class honours. By then he and Mimi had two sons, Mickey and Maurice. Their fourteen-year-old under-nurse, Agnes Moffitt, remembered the family well: 'They were a couple very much in love. She was a very dear person, and he was a wonderful man. He was born out of his time. He never made much of his title, and he always dressed a bit differently from what you would expect. One time I can remember him wearing a sombrero!'

'He was especially fond of children,' she continued. 'There were always wee urchins hanging around in the mews where he parked his car, near their house in Moray Place. One weekend he swept them all up, took them into town for tea, and bought them a big bag of toys at Woolworths.' Such gestures of *noblesse oblige* were characteristic of Suffolk, allowing him as they did to use his status and resources to circumvent good form and convention, and to indulge a conjuror's love of surprises.

Though a big man, and often animated, Suffolk had none of the overbearing traits of his mother. He stooped, and had a surgeon's nimble and sensitive hands, with a healing touch to them. Agnes recollected that if 'anyone on the staff was sick they would always go to him and he would sort them out'. This feeling for others also manifested itself in a talent for mimicry, an ability to assume other roles recalled by his professor at Edinburgh along with his 'charm of manner and the easy courtesy of the typical noble'.

Shortly before the war, the Suffolks were finally able to move into Charlton when his mother decamped to Somerset. Suffolk set up a laboratory there, and in 1939 he began work at the recently established Nuffield Institute of Medical Research in

Oxford. The outbreak of hostilities need not have put an end to
that, so it demonstrated a strong sense of duty that he applied to
rejoin his old regiment. To his chagrin, his application was turned
down because of his rheumatism. Feeling that he still had much to
offer, he began looking around for strings to pull.

The most obvious of these unfortunately looked as if he would
be of little help. Suffolk's cousin Cimmie – Curzon's daughter –
had until her death in 1933 been married to a politician with a
promising career, Oswald Mosley. His conversion to fascism, and
the friendships that he and his new wife Diana Mitford had made
in Germany, meant that he had lost any influence in government
circles. Suffolk turned instead to a friend of his father's, the 5th
Earl of Clanwilliam, who had been a fellow ADC to Curzon and
was now a Whip in the House of Lords. At the start of 1940, he
got Suffolk an interview at the Ministry of Supply.

Since he seemed to know about science, Suffolk was soon
directed to Gough's office. Gough was somewhat nonplussed at
being told to find an earl a job. 'I was frightened at being landed
with a titled member of Society about whom I knew so little,' he
said later. 'Such work as ours could carry no one who was not
there on technical merit. In that frame of mind I saw Suffolk.' Jack
Howard pleaded with him to be given a chance. 'His enthusiasm
was tremendous. Before long his infectious personality, his gay
buccaneering spirit won me completely.'

Gough had not yet become involved with bomb disposal, but
as Director of Scientific Research he had many other concerns.
One of these was liaison with his counterparts at France's Ministry
of Armaments. Suffolk's fluent French and his interest in engineering recommended him for the post as Gough's representative
in Paris.

At first this was to be a temporary arrangement, with Gough
clear that Suffolk was on probation, and unpaid. By February he
was installed at the Paris Ritz and busily acquainting himself with
a broad range of developments made by the French. A stream of
letters flowed back to London with news about everything from

mine detectors to anti-allergy treatments. They are written with puppyish energy and tailed with Suffolk's extravagant signature, all Elizabethan loops and strokes. For his part, Gough – or 'Master', as Suffolk called him – was charmed by his protégé's high spirits and a little flattered by the bond that was growing between them. After McColm, Suffolk had once again found someone who was almost a surrogate father. Gough fondly warned him not to over-extend himself and gently moderated some of his wilder schemes, reminding him that he was in France on government business.

One of his tasks was to guide Desmond Bernal around Paris when he visited it in April. He and the South African-born zoologist Solly Zuckerman had been conducting tests on the effects of bomb blasts on chimpanzees (as substitutes for humans), and the pair had come over to discuss similar experiments being done by the French. In particular, they were to confer with Colonel Paul Libessart, the Chief Engineer of the Artillery.

The British duo wanted to discover how close someone needed to be to an explosion to suffer injuries from the shock wave itself, rather than through incidental damage from debris. Libessart told them that his trials showed that the zone of highest pressure, at the centre of the blast, was only about twice the width of the bomb. This helped to confirm Bernal and Zuckerman's belief that there was much more chance of surviving an explosion than was commonly thought. Meanwhile, Suffolk took them to a succession of memorable parties and dinners. It was a most welcome change from a Britain where rationing was already in force; Zuckerman would soon be supplementing his food with meat from animals killed in his experiments.

Within a month, though, Paris was in a frenzy of fear at the German advance. At first, Suffolk wrote in his letters to Gough that his fellow liaison officer, Major Ardale Golding, was worrying unnecessarily with his talk of evacuating the mission. But by mid-June the two of them had decided that it would be sensible to move to Bordeaux, following the rump of France's government. They drove down slowly, past long lines of refugees and the shattered remnants of the French army. With them was Suffolk's twenty-seven-year-old secretary Beryl Morden, who had first worked for him at the Nuffield. Originally from Dunstable, she had got engaged at the start of the war to a fighter pilot, but after he was posted to Malaya she had loyally agreed to follow Suffolk to Paris.

At Bordeaux they found 'the most utter chaos'. The streets were crowded with rich and poor alike. Soldiers, sailors, financiers, ministers, former ministers and ordinary families had all fled from the Germans. There was no room left in any of the hotels, so for several nights Suffolk, Golding and their two secretaries slept in the car. Nor were any of the means of communication working, making it impossible for them to contact London. Proceeding 'upon our own initiative', they offered to carry to safety in Britain any scientists or engineers that Raoul Dautry, the Armaments Minister, could gather together.

Then the following evening, Sunday 18 June, the French Cabinet fell. Suffolk and Golding now found themselves met 'with nothing but a most obstructive and defeatist attitude from the higher members of the Bordeaux Government'. Eventually Suffolk got in to see France's new leader, the veteran Marshal Pétain. From him, following representations made 'in the most uncompromising and bald terms', Suffolk secured permission to embark the experts he had so far rounded up.

Among them was a physicist whose German accent had already drawn many suspicious glances in jumpy Bordeaux. Hans von Halban was born and educated in Leipzig, but he was of Jewish descent and had no love for the Nazis. For the last three years he had been working in Paris for the Nobel laureate Frédéric Joliot-Curie, but when the Germans invaded he entrusted von Halban with a special mission that sent him south. So began one of the most extraordinary episodes of the war.

What he and Joliot-Curie had been investigating was nuclear fission, and in particular the most efficient method of generating a chain reaction in uranium oxide. Central to this, it was realised, would be a moderating substance that would slow down the speed of neutrons being released from the uranium nucleus by fission so that any chain reaction became slow-burning and self-sustaining. The moderator must therefore absorb the kinetic energy of the neutrons without also absorbing the neutrons themselves.

If this splitting process could be mastered, the energy liberated from the mass of the nuclei would be enormous, as Einstein's most famous formula indicated. One use of it, as both French and German scientists had appreciated, would be to make a very powerful bomb.

The most promising moderator that Joliot-Curie had so far found was heavy water. This was water which contained high levels of deuterium, an isotope of hydrogen, and so was also known as deuterium oxide or D_2O. The world's only source of it in any quantity was the Norsk Hydro plant in Norway, and in February 1940 its manager had sent all he had to France in twenty-six two-gallon cans after learning of its potential military applications – and after refusing a German offer to buy it.

It was this that Joliot-Curie told von Halban to remove from Paris, together with documents about their research and some radium. 'I put my wife and one-year-old daughter in the front of the car,' he remembered, 'the one gramme of radium at the back and, in order to minimise any possible danger from radiation the cans of heavy water in between.' Eventually they arrived in Bordeaux, where von Halban decided to take the D_2O to England.

For several days Suffolk had been trying to find a French vessel willing to make the journey. Then the commercial attaché at the British Embassy came up trumps, directing them towards Captain Paulsen of the newly arrived *Broompark*, which had Scottish owners. He also consigned to Suffolk a package containing the small matter of three million pounds' worth of diamonds. The property of the Antwerp diamond bank, they had been spirited out of Amsterdam ahead of the Germans; today their value might be as much as four hundred million pounds.

Suffolk now began to load his cargo that, as well as scientific apparatus, included six hundred tons of valuable American machine tools he had found abandoned in wagons at the port. But his work had not gone unnoticed. 'By Tuesday evening,' he revealed, 'it was obvious that our presence in Bordeaux was known to the 5th Column, since there had been an attempt, which fortunately was unsuccessful, to bomb our ship.' By early the next morning, twenty-seven scientists and their families were also aboard, including von Halban and a second member of Joliot-Curie's team, Lew Kowarski.

Even after weighing anchor they were still not safe. The estuary had been mined, and another ship had been sent to the bottom that day. Knowing that the same might happen to them, Suffolk drew up a memorandum in French: '*faites en mer a bord du S/S* Broompark'. This he made von Halban and Kowarski sign, giving him permission to transport '*les biens*' – the goods – out of France without the usual formalities. It also allowed him to sink the raft now fixed to the bridge of the ship, on which was the heavy water in a dozen jerry cans, if it was in danger of being seized. Everyone kept a sharp look out for German destroyers.

The crossing was in fact uneventful, and on the morning of Friday 21 June they arrived at Falmouth. Suffolk telephoned to the MOS with news of his shipment, and with typical brio procured a special train to convey it and the scientists under armed guard from Cornwall to London. Early the next morning, Suffolk and Golding strode up the platform at Paddington to be greeted by a relieved and excited Gough.

Suffolk's news had thrown half of the scientific establishment into a spin. Gough had telephoned his junior minister, Harold Macmillan, at dawn to tell him of it, and after the precious heavy water had been sent with an escort to Wormwood Scrubs for safe keeping, Suffolk was summoned to the Adelphi to be debriefed.

He made quite an impression on Macmillan. In his memoirs, the future Prime Minister recalled 'a young man of somewhat battered appearance, unshaven, with haggard eyes' who combined 'charm and eccentricity'. He explained to Macmillan that he had brought over a large consignment of diamonds, some French scientists and 'something called heavy water'. Macmillan had no idea what heavy water was, and did not like to ask, but he was much taken by Suffolk's resourcefulness. It emerged, too, that there were other experts, and perhaps some uranium ore, in hiding at Bayonne, and Suffolk asked that arrangements be made to try to bring these across as well.

Never having forgotten that dramatic Saturday morning, Macmillan was to follow the Earl's later exploits in bomb disposal with interest. 'I have had the good fortune in my life to meet many gallant officers and brave men,' he wrote afterwards, 'but I have never known such a remarkable combination in a single man of courage, expert knowledge and indefinable charm.'

Suffolk was able to go down briefly to Wiltshire to see his young family before attending a meeting on the Tuesday at the Great Western Hotel, where many of the scientists had been lodged. By then, Halban and Kowarski had already had discussions about their work with Cockroft, Zuckerman, Bernal and Gough. They had disclosed their discoveries about chain reactions and heavy water, and though at this stage it was thought that any atomic bomb would not be 'explosive in the ordinary sense', it was clear that 'the radiations produced would be such as to render uninhabitable a very large area'.

Unknown to the French team, however, was the fact that a few months earlier the British had also begun work on fission in uranium. This top-secret project was steered by the Maud* Committee, which was composed of half a dozen scientists, including Cockroft and Blackett. At their discussion at the Great Western, 'my scientific friends', as Suffolk called them, were invited to join the British effort. Two weeks later the pair found themselves at the Cavendish Laboratory in Cambridge, where the project was housed. The research done there would ultimately

* Military Application of Uranium Detonation

form the basis for the successful Allied attempt at building a nuclear bomb.

In the meantime, Gough and the Ministry's senior military adviser, General Sir Maurice Taylor, had been racking their brains as to the safest place to store the deuterium oxide. Then Taylor had a very military brainwave. What was more impregnable than the King's own house? Off went a letter to Lord Wigram, the Deputy Constable of Windsor Castle, couched in the most opaque terms.

'I have a very unusual request to make,' wrote Taylor. Suffolk had rescued from France 'probably the most valuable and rare material in the world' which 'may prove to be without exaggeration the most important scientific contribution to our war effort'. Could Wigram meet Suffolk and think of 'some small chamber in the depths below Windsor Castle' where this substance could be secured?

By mid-July, twelve tins of 'special fluid' had been moved by Suffolk from Wormwood Scrubs to Windsor, and the help of the Castle's Librarian Owen Morshead had been enlisted. 'The King knows it is here,' Morshead informed Gough, 'and Lord Wigram knows . . . No one else at this end is, or will be, in the secret.

'But in case of extreme necessity, should I not be accessible (for Lord Wigram would not actually know how to get at it) I will give you another way round. There are two people here who would know how to get it if you told them that *it is in the same spot as the Crown Jewels*. They know nothing of this stuff: but they do know how to get at the Crown Jewels.' The tins would eventually be transferred to Lord Beaverbrook's new Ministry of Aircraft Production, which by late summer had won the ministerial battle to supervise atomic research.

No formal recognition was ever given to Suffolk's part in that. On 27 June, in a secret session of Parliament, Herbert Morrison related the outline of the mission, though he spoke only of valuable stores, 'some of them of almost incalculable scientific importance'. He also paid an anonymous tribute to the 'two officers of the Ministry' who had saved them. Beyond that, there was nothing to commemorate an adventure worthy of Drake or his ancestor Thomas Howard, one Gough thought 'brilliantly handled and crowned with conspicuous success'.

There was a private celebration, however, to welcome back the Ministry team and to welcome the Frenchmen to London after the meeting at the Great Western Hotel. At a supper party at a fashionable restaurant, Suffolk was on such scintillating form from late in the evening until early in the morning that the memory of it long stayed with Gough.

First a professor at the Sorbonne was treated to a dissertation on pharmacology in rapid French, interspersed with some choice bits of Paris gossip. Then, suddenly, in Suffolk's hands and gleaming in the candlelight were his treasured revolvers, produced from hidden holsters. Their merits were 'expounded with boyish delight'. Later the company was treated to a mesmerising dialogue between a Cockney speaking rhyming slang and a Chicago gangster conversing in his own argot.

The effect of such kaleidoscopic changes from the sublime to the ridiculous, fuelled by such inexhaustible energy, could sometimes be bewildering: Suffolk so full of ideas and so eager to entertain that he could seem almost brusque. Beneath that, as Gough saw, was someone who was also rather shy, and who compensated for that with noise and extravagant behaviour. Now, though, it was time for Wild Jack to rest a while. Then Herbert Gough had a vital new task for him.

6

SOMETHING WICKED . . .

On the day that Suffolk and his haul arrived in London, France signed an armistice with Germany. Britain now stood alone. Since the miracle at Dunkirk three weeks earlier, its people had been largely kept ignorant of all but the most sanitised version of events in France, but even so it was clear that things were bad. Although tens of thousands of troops had been rescued from the beaches, and many more taken off at St Valery in mid-June, losses of both men and equipment had been heavy. Invasion seemed inevitable, and morale was at rock bottom.

With the benefit of hindsight, historians might say that the situation was not as bleak as then believed. The Royal Navy presented a near-insurmountable obstacle to any crossing of the Channel by German troops. Britain's imperial possessions still willingly offered their aid. Churchill had newly become Prime Minister, and at his urging America was slowly beginning to stir. Yet at the time, even those best placed to see the big picture were glum. In late June, the Permanent Under-Secretary at the Foreign Office – Britain's senior diplomat, Sir Alexander Cadogan – surveyed the state of the war for his diary:

Report from A.E. [Anthony Eden] of a tour in Sussex and
Kent, which certainly makes it seem that the Germans can
take a penny steamer to the coast and stroll up to London! . . .
Certainly everything is as gloomy as can be. Probability is that
Hitler will attempt invasion in the next fortnight. As far as I can
see, we are, after years of leisurely preparation, completely
unprepared. We have simply got to die at our posts – a far
better fate than capitulating to Hitler as these damned Frogs
have done. But uncomfortable.

The German forces, however, would need time to recuperate
from their French campaign. The Luftwaffe in particular had suffered significant losses, not least from the Hurricane squadrons
stationed in France, and while bombing had begun of some strategic targets in Britain it was neither constant nor intense. Only
about two thousand high-explosive bombs were to be dropped on
the UK in all of June and July 1940. The attacks gradually crept
towards the capital. The first bomb to fall on the London region
landed outside Croydon on 18 June, while its first casualty – a
goat, at Colney – came a fortnight later.

By then the first 109 Bomb Disposal sections had been raised
and deployed. Among those that were busiest were Stuart Archer
and Edward Talbot's detachments in South Wales. After their
sobering introduction to the hazards of the work down at the
docks, Archer was getting to know his men. His NCOs all had
more experience than him, especially Serjeant Adams: 'He was a
signwriter by trade, and did some very nice work in that way, putting up a notice outside our HQ.

'He was a good sergeant. He would do exactly what he was
asked to do, and would get the right men to do the right thing.
He had a closer relationship with the men than I did, so was
better able to allocate the work. We were all doing a difficult job.'

The spirit of camaraderie that soon became characteristic of
these first, small sections was fostered not just by the common
dangers but also by the shared privations. 'We were very much in

our infancy,' Archer remembered. 'Nobody had even heard of a bomb disposal section. There was no provision for tools and there weren't any Army vehicles available, so at the start I drove a cattle truck.' The sappers poked their heads through its bars and mooed at surprised pedestrians as the lorry rattled through Cardiff.

Although the fledgling BD organisation had now been greatly amplified, it still had little experience of what was required for bomb disposal, nor were its needs regarded by the Army as a high priority. Accordingly, at a moment when everything was in short supply after the debacle in France, BD was again in for a period of neglect. This was also due to no hierarchy having been established for the new units, which were under regional control and expected to be largely self-directing. Administrative problems dogged all of the eight sections in Wales for several months.

'The demand for clothes and equipment to replace what had been abandoned in France was so great that we were given only a rifle, a spare pair of cellular underpants and two towels,' recalled Sapper James Lacey, who was at Swansea with 103 BD. Lacey had been invalided back from Belgium with tonsillitis shortly before the Blitzkrieg began. Bored in the Chatham barracks, he had been told that volunteers were needed for agricultural digging. 'With the thought of Kent apple orchards in the June sunshine, and the chance of getting out of those high walls, I had no hesitation – just what I wanted.' But instead of apple-scrumping, he and a dozen others found themselves on a train to Wales, with even the sergeant in charge unsure what the posting was.

Arriving late in the evening, they were forced to doss down in a YMCA. 'We were just settling down when a young officer came in ... and smiled at us and said "You're my new Bomb Disposal Section!"' Then there was an air raid.

The next morning the ablution and messing facilities made available were those usually provided to coalers on the docks. Next, the men were issued with picks and shovels and shown a hole where there was thought to be a UXB. There was no training. Gingerly, they used their hands and fingers to unearth the bomb and then took cover while the officer used his Crabtree to discharge it. Six more were dealt with in the same fashion.

That night an empty garage served as a dormitory, and 'bed was three blankets on a wooden floor'. Lacey became the section clerk, and a neighbouring Artillery unit showed him how to fill in the supply forms that would allow them to live within the Army. Transport – some motorcycles and an unreliable Bedford truck – were requisitioned off the street, and gradually other kit started to arrive. Nevertheless, the existence of Bomb Disposal remained a secret, and personnel had to use a civilian address to receive mail. One consolation was that for some time they were the only soldiers in the Army to be granted any leave.

Archer's experience of those early weeks was much the same. He had also managed to scrape up some basic tools, and with these he and Talbot set off to deal with their first UXB. This was in the middle of a shopping street on the outskirts of Cardiff, and was as novel an event for the locals as it was for the two BD sections.

'The wardens didn't know what to do about it,' Archer recalled. 'I arrived fairly quickly and said the place had to be evacuated for a radius of fifty yards. The police just knocked on the doors of shops and houses while we were digging and said "You have to go away". We were digging from 9 a.m. to 7 p.m., and eventually we got down to it. It was about 10ft down.

'The hole we dug was about 6 or 7ft square. We had to break up the road surface – that was the most difficult thing. The ordinary picks that we had were not much good, we needed hammers and chisels.'

Having found the UXB, they had little idea how best to proceed. So Archer took lives into his hands. 'We knew later that, in many cases, we shouldn't be disturbing the fuze at all. But here we didn't know any better.' The bomb was simply heaved up out of the hole by the section. Nothing happened, so the two officers tossed for the honour of disposing of it. Talbot won, and blithely humped it on to his shoulder before carrying it down to a beach near by. There he put a slab of gun cotton on to the bomb and blew it up. 'We didn't attempt to defuze it,' Archer remembered. 'We didn't have a clue.'

Looking back, he would appreciate that they had taken an enormous risk. 'I became aware afterwards that so many things I did really should have ended up with me going . . . There is no question that the Germans were ahead of the game.'

It would have been understandable – and natural – for the dangers of the work to start to prey on him. What prevailed instead was a sense that he must not let himself and others down. 'I think I accepted that this was just the job I was sent to the place to do, so you just get on with it. I am glad to say that I didn't suffer from nerves.' In time, and with repeated success, he became more confident but, as he said, it was always a matter of just taking the fuze out and hoping for the best.

What mattered almost more to him than his own feelings were those of his men. As in every other aspect of British life, the Army – and therefore Bomb Disposal – was divided along class lines. The men did the manual labour, and then would retreat to safety while the officer engaged the fuze in single combat. Similarly, the military reflected the paternalistic values of the landowner. The officer looked after the welfare of his section, and their behaviour reflected the confidence that he gave them. The sergeants and corporals might be the real heart of the squad, the ones who did much of the work, but the officer was still the team captain.

Therefore, what Archer was most proud of was 'that these young men who were my sappers and NCOs didn't suffer any nerves either, or at least they didn't make it known to me . . . They were just ordinary chaps who had been put into this work, without explanation. But they didn't complain.' By the autumn, those working on BD would be offered the chance to quit the work after six months of it, but very few took up the opportunity. None of Archer's men was among them, 'so presumably they dealt with their fears in the same way as I dealt with mine'.

The life of a BD officer in these months could be lonely. The sections were frequently quartered far from any base and the life of the mess, and the divisions between the ranks meant that any socialising between officers and men was often awkward. The latter had their muckers, but the officers might have no one with whom to talk and briefly to forget the strains of their work. Archer was fortunate to have both Talbot and his wife to relax with, not least because he disliked going to pubs.

Since all the sections were initially independent of each other, even those stationed in Swansea and Cardiff rarely met. It came as a welcome change when, in mid-July, 104 and 105 BD Sections joined 103 BD at the National Oil Refinery, Skewen, one of the principal targets for the increasingly bold Luftwaffe. Their latest raid had left thirteen UXBs.

Slowly the sections had begun through incidents such as this to teach themselves the essentials of their new trade. There were four phases to it: reconnaissance, recovery, defuzing and removal. In the first, a report of a UXB was followed up for confirmation. In many cases it would prove to be a false alarm, or a hole would turn out to have been caused by a bomb that had already exploded.

Should a UXB be suspected, then the officer would need to decide where to dig for it. This was not always obvious. The shape of a bomb and its momentum on impact sometimes took it deep into the earth, on a course shaped unpredictably by rocks or pipes underground. The path that it left could be probed with a long metal pole, but often there was nothing for it but to dig, and to dig deep. Bombs were starting to be found twenty-five or thirty feet down, meaning that the sapper now needed to become a skilled navvy, capable of sinking a shaft even in soft running soil.

In these early weeks there was little timber available for lining these shafts, so they were shored up with whatever came to hand, notably doors and floorboards from bombed-out buildings. Even so, accidents happened. Timbering took time and effort that was sometimes skimped, and waterlogged shafts crumbled too easily for safety. Tools were still inadequate – no more than spades at the start – while the only light available was that of nature. And there was the ever-present fear that any disturbance might set off the bomb.

Once it had been reached, defuzing seemed so far to present no great difficulty. The bombs had impact fuzes that had failed to operate, or if one still retained a charge it could easily be drained by the two-pin Crabtree discharger. Even so, as William Wells observed in Glasgow, 'We were rather dubious about this way of discharging the fuze for it was obvious that Jerry would soon be wise to what we were doing, and it would be a very simple matter to change the electrical connections in the fuze so that when the pins touched the socket the fuze would function, and the bomb blow up.' Soon enough, the Germans were to do exactly this.

Other features of their enemy were also becoming known to the BD squads. Most bombs dropped weighed fifty kilograms and were two and a half feet long, with a few of the larger 250-kilogram and five-foot long five hundred kilogram sizes also being found. They were constructed in three pieces, with the nose, body and base welded together. The tail fins were made of a lighter alloy attached by screws or rivets and were often torn loose on impact – a useful clue that a UXB was present.

About half the weight of the general-purpose or *Spreng Cylindrisch* (SC) ordnance was explosive, while the more stream-lined and thicker-walled *Spreng Dickenwand* (SD), made of one-piece drawn steel and used against reinforced targets, had an explosive content as low as 16 per cent. Once buried, the destructive power of the SD depended not so much on the quantity of explosive as on how close to the surface it lay, although that would make little difference to anyone right on top of it.

It was not just the British who were drawing lessons from these early forays. The Luftwaffe had also been busy assessing its performance since the beginning of the war.

Its chief problem during the winter after the first initial campaigns had been a shortfall of bombs. Hitler called for a vast increase in production, and the German genius for industry came into its own. The dimensions of bombs had always been fixed so as to conform to the standard sizes used by steel plants, and now compromises – such as allowing bombs to be welded together in three pieces rather than cast whole – enabled output to be quadrupled to a hundred thousand a month. Nonetheless, there were still difficulties with manufacturing enough explosive to fill them, and for a time Polish and Czechoslovak armaments were pressed into service.

Rheinmetall were told to simplify their fuzes in order to escalate the numbers made. The company had, however, already proved its worth, allowing bombs to be delivered pre-fuzed to airfields at the front, saving assembly time and manpower.

The Luftwaffe's discontent was focused elsewhere. One cause was that the lack of sufficient quantities of SD bombs meant that it was having to drop SCs on fortified objectives. This had led to some bombs shattering on impact, scattering their contents and fuzes before they had activated. More worrying, however, was that simply hitting the target was proving more difficult than anticipated.

The likes of H. G. Wells had imagined futuristic bomber fleets as not merely implacable but infallible. In fact, in 1940 precision bombing from altitude was a still far-from-perfect science. As with shooting, compensation needed to be made for the wind when aiming, but this was hard to gauge in practice, and more difficult still when under attack. A change in height by a pilot for a fraction of a second when travelling at 180mph could alter where a bomb landed by eight hundred feet. In daylight and in training, half of all German bombs could be delivered to within three hundred feet of a target, but by night, in the rain and flying higher to avoid anti-aircraft fire, the margin of error was far greater.

Near-vertical dive-bombing – in which the relative speed across the ground was much reduced, as was the height and time of the bomb's fall – did make for greater precision. But the Stuka only carried a small payload and thus could never be a strategic heavy bomber like the Dornier or Heinkel. Incendiary bombs that lit the target for these was one solution to poor accuracy, but they weighed little and were easily blown off course. Flares proved more reliable and, later on, large cities such as London and Coventry presented marks that were easily found. Even so, by then any pretence to precision strikes had largely been replaced by area bombing.

Certainly, by early August 1940 the Luftwaffe inspectorate's verdict on its bombing record was damning, not least because time and again many missiles were being lost through simple mistakes: '1. The bombs are dropped too low. 2. The wrong choice of fuze. 3. The wrong type of fuzing.' The last point referred to an especially common fault, as the BD squads knew: up to half of all bombs dropped at low-level failed to explode because they had not fallen for long enough to charge. Nevertheless, there was little cause for RAF Bomber Command to feel smug at its rival's shortcomings. The following year, a report would show that only one in three of its aircraft was getting within five miles of their targets.

On the morning of 18 August Lance-Serjeant Jack Button began to walk down the railway line from the station in the small Hampshire town of Hook. With him were half of 48 BD Section, due to relieve the men who, since dawn, had been digging for a suspected UXB in an embankment. There was some urgency to get the job done, as while they were at work no trains up or down from London could pass.

When Button arrived, the first excavation party had just reached the bomb. It had landed the day before, and provided they were careful it should have been simple enough to complete the job. The two crews changed over and, as was his prerogative, Button settled down to read the Sunday newspaper and have a smoke. He was familiar enough with a spade, having worked as a gravedigger before the war.

No sooner had the other soldiers left the site than they heard the bomb go off. Rushing back through the smoke, they were met by the bloodied form of Button. His tunic had been ripped open by the blast and he was dragging two injured men away from the crater. He himself had been thrown a great way, but he refused all attention until he had checked on everyone else. Five sappers and an RE driver, none of them older than twenty-three, had been killed.

A disaster like this had never before been suffered by BD. Indeed, its first-ever fatalities had occurred only three days earlier, at Shoreham, but worryingly the circumstances were similar. Two sappers, Frank Holman from Saltwood in Kent and Tom Smith from Beswick, Manchester, had been working on a UXB more than two days after it had dropped when it exploded without warning. Clearly the Germans had changed the rules of the game.

The trigger for this had been the start, in the previous week, of what would become known as the Battle of Britain. Although the Luftwaffe's main task was to eliminate the defensive capability of the RAF, in large part by bombing its airfields in southern England, the coming land and air assault also required the destruction of related targets. These included ports and dockyards, refineries and railways, aircraft factories and ordnance works, all of which came under greater attack. Then, on 13 August, Second Lieutenant Harry Mitchell saw something different about the fuze he had extracted from a 250-kilogram UXB that had fallen four hours previously. It was longer, it was marked with the number 17 on its boss instead of the standard 15, and it was ticking loudly. By two o'clock the next morning it was being taken to pieces by the Research Department at Woolwich. Later that day, a second fuze arrived there, this time recovered from an electricity depot at Lyndhurst, near Southampton.

What the technicians found was a fuze in two separate parts. The top worked like the Type 15 to which they had become accustomed, but when the igniter fired it did not detonate the bomb. Instead, it set off some thermite, the heat from which was conducted to the lower part of the fuze. There it melted a wax pellet, so allowing a spring-loaded plunger to rise and free the balance wheel of a clock, the mechanism of which then began to run.

It was clear that, using a key, this could be set by the armourer for a range of times, like an alarm clock, until eventually a slot in a disc aligned with a trigger and the striker was released. It was discovered that Mitchell's fuze would have gone off thirty eight hours and forty-five minutes after it had fallen, and that from Lyndhurst – which only began to run once it reached Woolwich – after just seven hours. Experiments soon showed that the delay period ranged from two to eighty hours.

Two other things were also clear. First, that the Crabtree discharger, the bomb disposer's talisman, was useless against this new fuze. The electrical part had done its work once the clock began to run, and this could not be stopped by belatedly tampering with the condensers.

Secondly and consequently, the ramifications of the arrival of the (17) were very wide. Its purpose was obvious enough. As General Marquard wrote after the war, the Germans had realised that a bomb with a long-delay 'would cause disruption to the enemy's armaments industry, to transport and the general public, and would tie up considerable bomb disposal resources'. Anywhere affected by such a fuze would need to be quarantined for at least half a week and it would detonate in any case. It sowed first disorder, then destruction. The chaos would be all the greater since every UXB would now need to be treated as a potential ticking (17) rather than as a probable dud (15), at least until it was uncovered. And as Button had learned, even that process had just become far more hazardous.

The advent of the (17) did not come as a total surprise. The possibility of a time fuze had been foreseen from the start by Merriman, and some had been dropped on British camps in France in May. Later that month, Gough had written to Paul Montel, the director of the Franco-British Scientific Mission, asking for more information. The details that Montel could give, however, were scanty: 'For the moment the working of these delay action fuzes is not well understood,' he replied, adding that their operation was thought to be chemical. When Paris was bombed by them in June, Eric Moxey was sent out to recover some examples but found the French authorities 'as usual . . . thoroughly obstructive'. He had managed to study a few and then, after an adventurous time he too escaped from Bordeaux.

The RAF, as was known, had their own long-delay pistols. Curiously, these also bore the number 17. They could be set to detonate up to thirty-six hours later, but there were recurring problems with their reliability and they were due to be replaced by a new chemically triggered series. One other point of potential importance was that they were protected as a matter of course by a booby trap.

The sections were alerted to the new threat, which led to a reconsideration of many aspects of bomb disposal practice. In Swansea, James Lacey's section had taken to storing in a garage the UXBs that they had dug up as they lacked both explosives and a site at which to finally dispose of them. One Saturday evening Lacey was alone on duty at their headquarters when his officer arrived with a lieutenant-colonel from the local anti-aircraft regiment. They wanted to have a look at one of the stored bombs, which had unusual markings.

The sergeant who had the only key was away so Lacey was told to break a window, crawl inside and undo the bolts on the garage door. Having examined the fuze, the officers decided that it would be safest to drop the bomb in the sea, so Lacey was made to manhandle its bulk on to the back seat of the colonel's Austin 7. Already apprehensive at being crammed in beside it as they set off, Lacey was made still less comfortable by the casual suggestion from the front, as the road became bumpy, that he sit on top of the bomb to keep it still. It was with some relief that he arrived in one piece at the dockside, and gratefully handed the UXB over to the Navy.

Lest it become too much of a strain on the sections, the War Office ruled early on that unless there were pressing reasons for dealing with a UXB, all should be left alone for four days after landing. Any with a live (17) inside would by then have blown themselves up. They were falling from perhaps two miles up and hitting the ground at some 545mph, and so the ability of clock-work fuzes to function afterwards was a tribute to the skill with which they were designed. The spindles and cogwheels were so light that they carried little momentum, while the balance wheel, which regulated the time, was clamped tight to the casing of the clock for protection.

The chance of a (17) not having exploded after eighty hours was therefore small but, as the Lyndhurst fuze had shown by only starting to tick once at Woolwich, there were always exceptions. It seemed to take little to free a mechanism that had jammed – even the shock of another bomb falling nearby would do it. So while it was sensible policy only to begin work on UXBs after four days, a party digging lustily down to a buried one would not know that their bomb had a (17) that must have stopped until they reached it. Moreover, there were going to be occasions when the damage that might be caused by a delayed-action bomb out-weighed the risk that it posed to life.

In the months ahead, there were to be only too many examples of this. One such occurred at the Hawker aircraft factory at Weybridge. On the morning of 21 September 1940 it was peppered by bombs, at least three of which were suspected of being (17)s. The first to offer help were members of a neighbouring Canadian Royal Engineers unit, who had no experience in bomb disposal. One of the UXBs had ripped through the factory roof and wall, and was embedded in a concrete driveway outside the assembly shed. The danger to the two dozen precious Hurricanes inside was obvious, as was that to anyone close to it. Without hesitating, Lieutenant John Patton and Captain D. W. Cunnington rolled the bomb on to a piece of corrugated iron and towed it away with a lorry before tip-ping the device into one of the craters created by the raid.

Two further fifty-kilogram UXBs had come to rest on the concrete floor of the workshop. One was speedily removed to a crater, with its fuze intact, by Lieutenant Charles Davies of a BD section that had arrived at the site. The other had penetrated the floor and needed to be dug out. The excavation began at once, and continued all night, despite the risk of the bomb exploding. Indeed, while this was going on the first bomb did go off outside. The digging operation was supervised by Davies, another veteran who had returned to the colours; he had been gassed while with the Welsh Guards in the First World War. At eight in the morning, ten and a half hours after the bomb had landed, it was pulled out. As thought, it was fitted with a (17) fuze.

For many weeks yet, there was to be no effective counter to a high-priority (17) except that of taking the chance of reaching and moving the bomb quickly, or of extracting the fuze. The latter was initially the preferred option, since the scientists needed specimens to study if they were to devise methods of defeating them – and means of making the soldiers' task safer. Yet the only way to get the fuzes needed was to take the very risks that made them so deadly. It was a quest for knowledge measured in human life, and one in which Archer and Talbot were soon caught up.

On 24 August UXBs had been reported near Loughor Station, west of Swansea, on the Great Western line that connected the industry of Llanelli with the ports of the coast. 103 BD's officer was in hospital with pneumonia, so Talbot had taken charge of the section. It took them twelve and a half hours to bring the first bomb to the surface, and when Talbot examined it he saw that it was a (17). Knowing the dangers, he decided to move it somewhere less susceptible to damage and, having ordered the men to take cover, hoisted the live bomb on to his shoulder. He then walked two hundred yards with it to a safer spot.

Meanwhile, another UXB had been sighted on a track near the railway. Sunday 25 August was spent digging in relays, with breakfast and lunch being provided by the local inhabitants. This time, when the bomb was uncovered it was found to have another new kind of fuze, numbered 38. A call was therefore made to the experts, Squadron Leader Eric Moxey and Captain P. W. Kennedy, the technical adviser at the War Office. Moxey proposed flying down straight away to try out his improved fuze extractor but this was vetoed by Kennedy, who perhaps had less faith in it than its inventor. He would come down himself by train the next day and try to extract the fuze so that it could be examined.

After arriving at Loughor he sent to Cardiff for one of the new steam sterilising plants, which were still in development. A Navy lieutenant who had seen it demonstrated at Manby brought up the equipment. Lacey was among those who watched for two hours as a disc-shaped hole was trepanned in the skin of the bomb. There seemed to be a lack of pressure in the system, and much of the steam was escaping from the nozzle.

The next morning, Archer arrived to have a look. The fuze appeared to be stuck in the pocket and a booby trap was suspected. Kennedy injected an experimental solution of ether-acetone into it, but this failed to do the trick. Eventually Merriman himself arrived towards evening and, from behind cover and using a long string, pulled out the fuze. Nothing happened. Merriman then pointed out that the switch on the steriliser was set halfway between steaming and trepanning, leading to the problems experienced with the kit. Only by overcoming such teething troubles could progress be made, but it was no wonder that BD officers sometimes felt themselves to be, as one put it, 'piggy in the middle, with the German boffins inventing things to kill us and the British boffins designing something that would enable us to get the fuze out without blowing ourselves up'.

The Loughor (38) had been ruined for experimental purposes, but the next day another was extracted intact by Archer at Port Talbot. They turned out to be standard electrical fuzes with a slight delay, intended for use against naval targets. Three weeks after his stroll with the bomb at Loughor, Talbot read in *The Times* that he had been awarded what was then the highest honour for courage shown by soldiers outside the confines of the battlefield, the Empire Gallantry Medal. The citation praised his 'devotion to duty'. Four awards had been made, one of them to Jack Button, the first public recognition of what was still a largely unknown and unseen contribution to the battle for Britain.

7

. . . THIS WAY COMES

Although BD work was accompanied by little fanfare, it was gradually becoming a more common sight in vulnerable areas such as South Wales. Like Talbot, Stuart Archer had been kept busy. For the first month or so he had been doing 'a lot of donkey work around the coast', and the section had started to become rather blithe about their lot. Once a fuze was discharged the bomb was completely inert. 'They were safe as houses,' he remembered, with a certain unintended irony. 'We took no notice of them, they were up and out and that was that.' Indeed, some of the sappers 'to be very naughty, would put one under someone's bed'. They were about to get a reminder of the frailty of the thread on which their lives hung.

St Athan aerodrome lies to the west of Cardiff, between the capital and Bridgend. On 15 July 1940, four 250-kilogram UXBs were dropped on the field, two of them within ten yards of the hangars where Spitfires were parked. Archer arrived at midnight, and by four o'clock the first had been dug up from about five feet down. He applied the discharger in the usual way, but when he then tried to extract the fuze he thought that he felt a spring behind it. He suspected a booby trap, and that the fuze might still be live. The only solution was to move the bomb.

As gently as possible, it was loaded on to a lorry. Archer himself drove this for two miles away from the airfield before the bomb was taken off and detonated with a slab of gun cotton. Then he returned to St Athan and dealt with the second bomb in exactly the same way. As he admitted, 'this was really taking a hell of a chance. But then you had to take a chance, otherwise you were not doing the job you had been sent to do.' Not everyone, perhaps, would have seen it that way, but that almost without exception such chances were taken speaks eloquently of the calibre of those who served in BD.

A month later, at Moulton, north-west of Barry, Archer was to have his closest shave yet. Having, as he thought, discharged the fuze in a 250-kilogram bomb, he found that the impact had distorted it so much that he could not get it out. 'It sounds a bit bloody silly, but it was jammed in the fuze pocket.' It was then that he made what should have been a fatal mistake. Many BD officers regarded extracting fuzes as a badge of honour, and having to blow up a bomb in situ as a defeat. It was when the red mist of frustration descended, as in cases such as this, that accidents happened. Refusing to be beaten, Archer grabbed a couple of pickheads and, using them like a pair of giant tweezers, began to prise out the fuze.

'If you imagine a pickaxe has a curve to it,' he recalled, 'I would put that in the groove on the fuze and would get leverage.' To his horror, when he finally yanked it out he saw that the number on the head was not 15 but 50. It was yet another new type and, as the analysts were soon to confirm, perhaps the most lethal to date.

The (50) was christened by Rheinmetall the *Sonderzünder* – the special fuze. It was similar in principle to the (15), but designed with a particularly high resistance so that the arming time was increased to about three minutes. This meant that it only became live once it had settled in the ground. Inside were three sets of trembler contacts, two being spiral springs surrounded by a narrow metal tube, the other a wire with two small balls balanced on it. This had a vertical axis. That of the other two was horizontal. This made the bomb sensitive to movement in any direction. Tests showed that a movement of one two-hundredth of an inch in one four-hundredth of a second by any of the tremblers might be enough to trigger an explosion. For good measure, if the charging plunger was depressed – for instance by the Crabtree – then the firing condenser was connected straight to the igniter.

It seemed that every time BD had worked out how to deal with a fuze, Rheinmetall would unveil a new type. It was becoming crucial to take note of the markings on the fuzehead, especially the identification number circled on it. In good German fashion, these numbers appeared to be standardised. For example, those in which the integer was five were always general purpose impact fuzes, such as the (15) and the new (25), which Archer had recently encountered.

BD supposed that the Germans avoided using random or false markings so as to help their armourers. For instance, it would be important to know for certain with which fuze a bomb was armed if an aircraft returned to base with a primed load. Yet while it was vital for the British squads to know what kind of fuze a number denoted, it did not stop the German scientists ringing the changes. Each new type was a conundrum that needed to be solved as fast as possibly by Herbert Gough and his team. They had, for instance, just discovered that if the (25) was discharged with a Crabtree it would explode.

The function of the (50) that Archer had found was unambiguous. If it were to be combined with a (17), that would prevent absolutely a delayed-action bomb being moved, whatever risks an officer might otherwise take. For the moment, the use of the two fuzes together lay in the future, and Archer was simply glad still to be alive. For some unknown reason, the tremblers in the (50) had not reacted to his brutal treatment of the fuze and he had got away with it: 'It should have gone off. It was just luck, luck, luck. A lot of my friends would probably do the same thing as me and be blown up. I was lucky, and they weren't.'

Although these developments made life considerably more difficult for Bomb Disposal, the attention of the rest of Britain was riveted on the skies above. There the RAF was holding its own, and as yet there had been few attacks on civilian centres away from the coasts. A survey revealed that one in three people in Notting Hill made tea when the siren sounded, and it was not until 16 August that there were casualties in London, when sixteen people were killed and fifty-nine injured in a raid that hit Wimbledon. A week later, the first bombs fell on the centre of the capital, largely in the East End. Perhaps it was the proximity of these that now stirred the politicians into action.

The day after that raid, 25 August, Churchill fired off an urgent memo to the War Office. He had noted that a number of the new 'delay action bombs' had been 'thrown last night into the City causing obstruction' and fancied that the Germans might even be so bold as to 'try them on Whitehall!'.

Now his long experience came into play. He well remembered the problems caused during the First World War by the Germans having skilfully used time fuzes to render 'impossible the use of railway lines'. This new menace was not to be under-estimated and it seemed to him that 'an energetic effort should be made to provide sufficient squads to deal with this form of attack in the large centres'. He appreciated as well that the work was 'highly dangerous' and 'particularly honourable', and that it should be encouraged by appropriate recognition.

The memo was to prove a turning point for BD. Until then, it had been the poor relation in the Army, a situation exacerbated in particular by the absence of any workable chain of command above regional level. Though the Director of Anti-Aircraft and Coast Defence had done his best to liaise with the civilian author-ities, bomb disposal was essentially a field operation and its technical and administrative needs could not be met from a desk in Whitehall. With the Prime Minister taking a personal interest in BD, its standing was transformed literally overnight.

The very next day, Lieutenant-General Bernard Paget, Chief of Staff to the Commander-in-Chief Home Forces, set out his thoughts in another memorandum. Only the week before, he stated, there had been much discussion about a reorganisation of BD, and in fact command of it had now been transferred from the War Office to Home Forces. There had also been a start made on doubling the number of RE sections to 220. The main obstacle to their greater effectiveness, however, was that the Police and ARP wardens were frequently ignorant of BD's existence, and the squads were often held up by them when trying to reach a UXB. At a meeting that week of the Chiefs of Staff Committee, chaired by Field Marshal Sir John Dill, the head of the Army, Paget suggested that better guidance be given as to which bombs should be dealt with urgently and which could be left until later.

Dill made clear his opinion of the work that those in charge had put in so far, and so that afternoon an emergency conference was held of the three Services and the government departments concerned, notably the MOHS. The Navy and RAF formally accepted responsibility for bombs on their own patches, and proposals to overhaul the organisation of the Army sections were discussed. A central clearing house was to be set up to improve the collection, exchange and distribution by the Forces of technical information, which had been poor, and an order of priority was to be set for the disposal of bombs, headed by transport links and oil storage depots. It was also decided that it should be the task of the regional Civil Defence headquarters to report UXBs to the military and to determine the urgency of defuzing them, and that BD were to deal with unexploded anti-aircraft shells.

The cumulative effect of these changes would be to put BD in far better shape to weather the approaching storm but, given the certainty of its arrival, they had been too slow in coming. Most would also take time to implement. The main improvement, however, was immediate: namely that due to Churchill's intervention the Civil Service was ordered to make the supply of equipment to BD a matter of prime concern.

Until now, it had always been at the back of the queue. Again, this was perhaps not surprising given that even by the end of August UXBs had largely yet to make their presence felt in London (although half of the 642 reported so far had fallen only a short commute away, in and around Tunbridge Wells). Greater priority had been given to the claims of the hard-pressed aviation ministries, and of the hard-pressing Beaverbrook.

A second important advance was the appointment by the Army of a Director of Bomb Disposal (DBD). Major-General Brian Taylor was, at fifty-three, winding down his career, which had begun as a cadet in the shadow of Queen Victoria's reign. Having seen service as a Royal Engineer everywhere from Bulgaria to Iraq, he had been about to retire when, in May 1940, he was made Inspector of Fortifications (IF) and ordered to supervise the building of pillboxes across Britain. No doubt he was given the additional appointment of DBD as the IF's area of responsibility dealt with a German invasion. The new directorate became known as 'Ifs and Buts' to those whose lives depended on the technical bulletins it issued, but in fact Taylor and his small staff did a superb job of remedying many of the earlier deficiencies of BD's structure. Information on new fuzes was gathered, analysed and redistributed much more efficiently, and the independent sections finally acquired a chain of command that could be of some help to them. Above all, Taylor managed – after some delicate early exchanges – to establish a harmonious and fruitful partnership with Gough.

Demarcation of their respective territories might have become an awkward problem, but both quickly agreed that they should cooperate as equals. Gough and the UXBC would confine themselves to 'matters of equipment and research' for the troops, while Taylor would pass on to him the disposal squads' requirements and their experiences of using in the field the apparatus developed by Gough.

It was simple enough for the Navy and the Air Force to agree to handle their own bomb-disposal work. They covered much smaller and better-defined territories than the Army, and since ports and airstrips had from the first been in the front-line, their organisations had also evolved more rapidly to deal with the threat. Many of the Navy's initial encounters had been with magnetic sea mines, which had been dropped in estuaries and harbours and had drifted ashore. The first of these had been recovered in November 1939 at Shoeburyness, by a team from HMS *Vernon*, the Portsmouth mine school that was to become the centre of naval BD activity. The scope of this was soon extended to include conventional bombs that fell on RN bases.

By the late summer of 1940 the RAF BD set-up had also been refined. The early X Station scratch squads were replaced with dedicated Bomb Disposal Flights. These consisted of about twenty airmen, commanded by an officer, but much of the burden of the defuzing fell on those senior NCOs who were trained armourers. When the Battle of Britain was at its height, there were 188 of these specialists at eighty airfields, and other mobile teams were held in reserve. With the Luftwaffe aiming to disable fighter stations such as Kenley and Manston, the flights' workload was heavy. In the last week of August, for instance, sixty-two UXBs were awaiting disposal at Biggin Hill alone.

The skill and speed with which the parties worked made a vital contribution to the Air Force's resilience in those strenuous months. Yet since they were controlled from the Air Ministry they were often overlooked for praise and awards by the RAF's leadership, despite the heroic efforts of some armourers and officers. In January 1941 Flight Lieutenant Wilson Charlton, who was responsible for clearing bombs from runways in much of the West of England, was awarded the George Cross for his work in September and October 1940. In the course of two months he had made safe more than two hundred UXBs, working day and night 'with undaunted and unfailing courage'.

Eric Moxey was among those directing technical operations from the Air Ministry, although he also found time to perfect his fuze extractor. His civilian job was in engineering – he had built the longest conveyer belt in Britain for the Austin works at Longbridge – and following the UXBC's recommendations he had by July designed with Merriman a lighter and stronger model of his invention, known as 'Freddy'.

Difficulties with getting it into service quickly, however, were starting to frustrate Gough. The Sparklet bulb gave the operator two minutes to get clear before extraction began, but the pressure induced by it, at 750psi, required modification as it was unexpectedly high. Then the manufacturers failed to make their delivery date, blaming their suppliers for faulty castings, so that by mid-August just ten were being trialled by RAF and Army units.

There were also promising developments with several of the other methods that had been discussed by the UXBC, most notably the steam apparatus that Talbot and Archer had seen at Loughor. Much of the work on this was now being done in Manchester by Metropolitan-Vickers, the electrical engineering firm that would build many of the Lancaster bombers, and it was due to go into production in late August.

So it was all the more galling for Gough to learn at the end of that month that, despite the assurances he had been given on taking over all responsibility for BD research, others were still pursuing separate lines of enquiry. And more irritating still, this was happening right under his nose.

The culprit was none other than the Research Department at Woolwich, whose experiments were now supposed to be overseen by Gough and the MOS. The RD was clinging to its independence, however, and with its roots in the Services seemed still to be giving its allegiance to its former master, the Ordnance Board. This was the tri-service body responsible for the Forces' munitions which historically had dictated much of Woolwich's work.

On 26 August, Edward Andrade told Gough that he had just been to a meeting of the Board at which were also present Moxey, Captain Kennedy of the War Office, and Dr Robert Ferguson from Woolwich. Much of the discussion had been about the newly discovered Type 17 fuze, and Ferguson was asked to see whether a liquid could be induced in to it to jam the clockwork mechanism. This was an extension of an idea first raised by Andrade at the UXBC meeting in May. Furthermore, it was clear that RD were also collaborating with the Royal Aircraft Establishment on the development of other techniques, such as using magnetism to interfere with the (17)'s mechanism.

Not only did Gough feel that this undermined his authority, but he also appreciated that it would lead to exactly the wasteful duplication of effort and resources that the UXBC had been established to avoid. The danger was that the scientists would be working for two masters, and taking twice as long to find the answer to the same question being asked by both the UXBC and the Ordnance Board. Gough had suspected that something was afoot when RD had failed to invite him to a meeting about (17)s earlier in the week. Then when his opposite number at the Ministry of Aircraft Production rang to complain that Woolwich was poaching his fuze experts for their work, quite without Gough's knowledge, he decided that he had had enough.

Gough wrote to the President of the Board, Air Commodore Huskinson, to point out that the MOS was not solely concerned with production requirements of apparatus for bomb disposal, as seemed to be thought, but also with directing research into it. Close coordination of this was, he believed, essential at that time, when any overlap of work might slow down the discovery of solutions to the new types of fuzes being found.

Gough's reasoning and stance were entirely correct, but this was a turf war that had years yet to run, and the day after the Ordnance Board meeting it claimed its first victim. The campaign against the fuze could not be won without samples for the scientists to study, and the more organisations that were involved, the more specimens were needed. Eric Moxey was under pressure from all sides to recover them. From the Board he knew that they were wanted by RD and by the research section of the RAF. The UXBC was developing a thirst as well, not least to supply Andrade's experiments. He also needed to keep testing Freddy on them.

On the afternoon of 27 August, in the Wolseley that he had hand-built himself, Moxey drove down through the Kentish lanes to Biggin Hill. Half a dozen raids had hit it in recent days, and had largely been dealt with by Warrant Officer Edmund Hunt, who met him at the field. Two five-hundred-kilogram bombs were embedded in the strip, and they appeared to be fitted with the new long-delay fuze. The Ordnance Board wanted them retrieved, and Moxey had volunteered to do so.

He managed to remove the first fuze without any trouble. The second, however, seemed less straightforward. He and Hunt thought that there might be some kind of anti-interference device connected to it, but as they continued to work on the bomb it exploded. Moxey died on the spot. Hunt was blown almost seventy feet by the force of the blast, but survived. He was able to collect himself sufficiently to help with clearing the rest of the airfield, yet such were the dangers of bomb disposal that his reprieve was only brief. Ten days later he was to be killed by another UXB.

In Moxey's Wolseley was a drum kit that he had bought to play with his son Jack. Instead, his father's death meant that Jack and his twin brother Bill had to be taken away from school at Malvern because their mother, May, was left with little money. Then, two years to the day that Eric Moxey was killed, their older brother Nigel died in an accident in Egypt, while also serving in the RAF. He was twenty-one. Remarkably, instead of letting herself be embittered by this series of tragedies, May spent much of the rest of the war nursing German prisoners.

The next time that the Moxey family saw the Wolseley was on VE Day, when it was being driven slowly down Whitehall. Sitting in it, waving at the cheering throng around him, was Winston Churchill.

The sky was on fire. Early on the morning of 2 September, Stuart Archer was telephoned in Cardiff and told to go as fast as he could to Swansea. The town had been heavily bombed a few hours earlier. The railway station had sustained much damage, and four wheat warehouses set alight. Eight thousand tons of grain was going up in flames. Incendiaries had also fallen on the ICI factory at Upper Bank. There had been many casualties.

Out to the east, the raiders had struck at the vast tank farms belonging to the Navy and the National Oil Refineries. The latter, owned by Anglo-Persian (now BP), had been Britain's first modern refinery when it opened at Llandarcy in 1921, and was still the largest in the country. It was a vital link in the nation's infrastructure, supplying fuel for everything from the merchant shipping fleet to London's buses. It was also one of the Luftwaffe's key targets. While U-boats prowled the Atlantic, preying on the few tankers still braving the voyage from Nova Scotia to Swansea Bay, the bombers struck at the fuel oil, petroleum and aviation fuel already stockpiled.

The dual attack was proving effective. In September, imports of oil fell by a third. Although splinter-proof walls were constructed around tanks, more than 76,500 tons was lost to bombing. This represented an eighth of Britain's monthly needs, so it was crucial that further harm be prevented if possible. Any UXBs near refineries were to be dealt with immediately, whatever the risks.

Even twenty miles away, Archer and his section could see the flames and smoke belching from Llandarcy. When they arrived they realised that half a dozen of the giant tanks were on fire. Each held some three million gallons of oil. The heat was extraordinary.

Any attempts to extinguish the blaze could not begin, however, until another danger had been resolved. A stick of four 250-kilogram bombs had been dropped on the refinery, landing scattered among the tanks, about two hundred yards from one other. None had gone off, but it was apparent that they were probably (17)s and might detonate at any time. The longer they kept the firemen at bay, the more fuel burned. It was a classic application of a long-delay fuze.

Archer inspected the four UXBs and decided the top priority was that which had lodged right under one tank, having first smashed its way through the concrete plinth on which this stood. There was a fire raging out of control only fifty yards away. Very carefully, the men began digging, working as they always did in relays of two for fifteen minutes. This minimised the time that each man was exposed to risk. Archer, however, remained by the tank all the time, encouraging his men.

The excavation was hard going, with the aim being to widen the path made by the missile so that Archer could reach it. Then, after an hour, one of the other UXBs exploded. A while later, so did the other two.

It was now two o'clock in the afternoon: eight hours since the raid. Conscious all the time that their bomb would be the next to explode, the section worked on amid the roar and smoke from the flames. Finally, the hole was large enough to admit Archer. The men lined it with some planks and he slid down head first, arms outstretched, into the dank half darkness.

The bomb was several feet down and clammy where water had seeped in. Reaching around it he could see that such had been the impact that it had cracked open, and that the face of the fuze had been torn off, making it impossible to identify. It would also be impossible to pull it out.

There seemed nothing to do but to accept defeat and let it eventually destroy the tank. Yet after so much effort that would rankle with Archer and the sappers. 'So then,' he said later, 'I did the most unusual and strange thing.'

He knew that bombs were filled with explosive via a hole in the base that was then closed off. If he could not remove the fuze, perhaps he could reduce the force of the blast by scraping out the charge. He therefore unscrewed the base plate and, finding that the explosive inside was in powdered form, began to dig it out with a trowel. For half an hour he went steadily on, alone and very aware that the fuze a few inches away might be ticking down. He could do nothing to halt it. He could only hope that his luck would hold.

Having emptied the back of the bomb, Archer saw that the tubular fuze pocket that stretched across the inside of the casing had sheared away at one end. 'So I grabbed hold of this, and with brute force and bloody ignorance I lugged it back and forth, back and forth, and I managed to get out of the back the whole of the fuze pocket, complete with its fuze – a clockwork fuze – and all the booster charges. God willing, I'd got it, we were there, we'd done it.'

It was about four o'clock when Archer crawled back up into the light clutching his prize. Eager to see what he had got, he took a pair of pliers and begun to tug out the innards of the fuze. Then he heard a sharp crack and saw a flash. It was as if something had gone off under the (17). Alarmed and puzzled, he pulled out the rest of it and removed the gaine.

'I looked inside the pocket, and thought, What the hell is this?' Behind the (17) was a second mechanism, which now slid into view. Archer was the first to uncover the secret anti-interference device that had already claimed several lives.

This was the *Zussatzünder* (auxiliary fuze) 40. Archer's find was rushed up to the War Office, where examination showed that it worked on very simple principles. As the (17) was inserted into the pocket, the gaine underneath it slotted into a second fuze, the Zus. Inside this was a needle pressed forward by a spring but held away from a detonator cap by a detent, a small rod balanced on a ball.

The impact with the ground shook the rod away, so that the needle was restrained just by an arm pressed against it by the gaine. The Zus was now set. When the gaine was pulled upwards – for example if the fuze was extracted – two knife edges gripped the pocket and ensured that the (17) and the Zus were separated. After a half-inch of movement there was nothing to hold the arm in place, and the needle was forced into the Zus's detonator.

Both Archer and BD had long been expecting the advent of such a device, and others had recently had close calls. On the day before Moxey's death, at Croydon Golf Course, Lieutenant Wallace Andrews had made several attempts to remove a fuze that seemed to be sticking. Thinking that there might be a trap, he had finally pulled it out from some way off using a piece of cord, but he had still been hurled a considerable distance by the subsequent explosion.

Meanwhile, at Great Livermere in Suffolk 2nd Lieutenant John Emlyn Jones had managed to get out another (17) twinned with a Zus, confirming BD's suspicions that in the space of a few weeks their task had become infinitely harder. They had had the measure of simple impact fuzes, but now sections had to contend with time bombs. Coupling them with the motion-sensitive (50) ensured that they could not be moved from important sites. Nor, now, could their delay fuzes be withdrawn without triggering the Zus. Protected by its two bodyguards, the (17) seemed invulnerable. Dropped in sufficient numbers, there would be nothing to stop it from paralysing Britain.

Stuart Archer knew that he had had a lucky escape. Two, in fact. Not only had the Zus not exploded when he took out the (17) – probably because some water had got into it – but, just as miraculously, the UXB he had chosen to tackle was the only one of the stick that did not go off at the refinery. Its clock ran down two hours later, once he was back in Cardiff. Had the bomb been down a little deeper, had the cogs turned slightly faster, he would be dead.

He was sustained in his work not simply by his deeply felt sense of duty, but also by a perverse comfort in the certainty that if anything went wrong he would know little about it. 'When you are sitting on top of a bomb,' he thought, 'if it goes off, then you're gone . . . You weren't going to be wounded and lying in agony. You were either going to be alive, or you weren't.'

Yet while in the next months he would voluntarily risk his life with a UXB some two hundred times, he did not regard himself as brave. 'I take my hat off to people who were running into enemy fire. That's the thing I would have been afraid of.

'The people that I was proud of were the young men who were working for me. They were not hand-picked. They were just butchers, bakers and candlestick-makers. But they were prepared to follow what I was doing, without any questions. That, I think, was courage.'

Not everyone, however, bore the strain so cheerfully. He and Kit and the baby had taken a flat in Cardiff, across the river from Bute Park. 'We had a young officer staying with us, and my wife noticed that he was getting very, very agitated indeed at the possibility of next being sent out on a job.

'She said to me: "For goodness sake, you do it, because he's absolutely scared to death." He was a brand-new man. I'm not blaming him, but obviously he wasn't suited mentally to the task.'

Stuart Archer was more aware of his wife's feelings. He never kept anything back from her, including the fact that at first he lacked the proper tools for the work. She also supported his decision to remain in BD when, after six months, he could have asked to be transferred. Yet inevitably it was a rather fraught way of life, with Kit in constant dread of hearing an unusually loud bang or an official-sounding knock at the door.

'It was extremely difficult for her,' confirmed Archer. 'Especially early on, when even I didn't know what was happening. She was with me all the time, never knowing if I was going to come back from a job.

'That was a bloody sight worse than what I had to put up with. I don't know how she coped.'

8

RED SKY AT NIGHT

London had never seen a night like it. The fires at the docks were so fierce that they blistered paint on boats three hundred yards away, hastening down the Thames. From warehouses gushed a lava of molten stock, coating the cobbles with seared sugar and charred rubber. The lungs of firemen smarted from pepper that had spilled from casks and filled the air. As they struggled to bring a blaze under control, embers a foot long would be hurled by the heat into the streets to set new conflagrations. Piles of timber that had been doused as a precaution began to steam just minutes later, and then flowered flame. Even the wooden blocks that made up the roads burned. The world was alight: the bombers had come.

That Saturday afternoon in September had been the most glorious of late summer. Then, towards teatime the heavy throb of engines had been heard to the south. Up the river they came – Beckton, Woolwich, Milwall, Rotherhithe, Limehouse – a tsunami a thousand aircraft long, raining down thunder on power stations, gas works and, above all, the docks. Behind them they left pulverised the narrow streets and cheap houses of East London as they swept on, high over Tower Bridge, until the wave spent its last fury on Westminster and Kensington. Two hours later they returned.

Guided by the fires lit by the first armada, 250 more bombers pounded the capital's hubs of trade and storage until dawn. On both banks of the river buildings blazed, turning the Thames the colour of blood. Death rode on flying brick and thudding stone, collapsing ceilings and falling walls, bright lances of glass. More than a thousand fires were started. By daybreak on Sunday 8 September 1940, 430 people had been killed and more than a thousand wounded. Dozens of streets were impassable, thousands of homes lay shattered. At half-past seven in the evening, the sirens sounded again.

That night, the City and East End once more bore the weight of the attack. Telephone exchanges, hospitals, and gas and water mains were hit, and all the railways to the south immobilised. More than four hundred died and 750 were hurt. The following evening, 370 lost their lives and more than 1400 were injured. So it continued every night for fifty-seven nights and then, after the briefest of respites, for week after week until May.

The instinctive reaction to the bombing of most people was fear, and a feeling of powerlessness. 'Most of us were scared stiff in the first week,' wrote one woman to a correspondent in America. 'We didn't show it much but we damned well were. Then I suppose the "soldiers of the front line" business came over us and we found that there was no point being scared, and also it was a bad thing to be, so we stopped being.'

In Poplar, Vi Regan, the wife of a bricklayer, spent the first afternoon of the Blitz crouched in the Anderson shelter in their garden. Her husband Bill had dashed off to Millwall for duty with a rescue squad and she was left alone listening to the crump of the bombs and the screams of her neighbours' terrified children. Her own daughters had already been evacuated to Oxfordshire. After a few hours she heard the warden shouting to people to leave the area, as some oil tanks were in danger of exploding. Vi helped to carry one of next-door's children up the street towards the safety of a school.

'Great canopies of billowing smoke blotted out the sun, and barrage balloons were falling down in flames,' she recalled. 'We were literally surrounded by terrible fires and the air was too hot to breathe.' Even once they reached the school, Vi was still frightened by the 'fierce crackle of the flames and the constant crashing of falling masonry, so near us, dear Lord. So near.'

When one looks at the photographs printed in the newpapers at the time of East Enders smiling cheerfully and giving the thumbs-up, it is easy to dismiss such images as propaganda. In fact, the transformation in people's response to the bombs from terror to resolve was rapid, as Celia Fremlin saw at a shelter in Stepney, which was hit heavily in the initial attacks. At first, 'they were screaming and saying "I can't stand it, I'm going to die, I can't stand it!" Sometimes the women would be hysterical, crying and falling on the floor.'

When she returned four nights later, the same people had already become accustomed to the sounds of the Blitz. Stools had been brought in and the atmosphere was far calmer. There was even a sing-along, 'because once you've gone through three nights of bombing and come out alive, you can't help feeling safe the fourth time'. The familiar streets outside might be being demolished, but by remaining themselves, and by holding on to their values, communities found a way of coping.

The writer of the letters to America was working in the East End too, for the Women's Voluntary Service. In the shelter in which she sought refuge there was 'no talk of horrors at all, the air raid wardens, in bedroom slippers mostly, went and fetched beer from the pub, which was miraculously still intact, tea and water . . . All the week I did this work, and the more I saw of those people the more I wondered. I've always loved them and got on with them, but I've never known just how grand they were before.'

Keeping one's feelings under control did not mean that fears went away. People clung to routines to get them through, but sometimes it all became too much to bear. One young woman survived unscathed the bombing of a restaurant. The next day she went to work: 'It was not till later in the morning that the tension broke. I was standing in my chief's office with some papers he had called for when I suddenly started to cry. The pent-up emotion burst and knew no bonds, my tears were uncontrollable. I was sent home . . .'

After a year of war Britain had begun to be bombed in earnest. With the RAF unbowed, the focus of the German strategy changed to destroying the nation's industry and infrastructure, but the chief target was London. The Luftwaffe defined a major attack as a raid in which more than a hundred tons were dropped, and only on the capital did more than ten tons ever fall in one raid in September and October. In London were concentrated the economic interests that could keep Britain in the war, and while the Nazis' planners did not expect the country's morale to crack unless the bombing was prolonged, they believed that severe damage to its sources of wealth would force the ruling classes to sue for peace or bring about a change of government.

The Port of London was as a result to become the single-most bombed target of the war. More than one thousand missiles fell on it, and a third of its sheds and warehouses were destroyed. More than fifty thousand high-explosive bombs were to fall on the rest of the city by the following summer, and by the end of September alone 5730 people had died there and some ten thousand had been badly hurt. Four out of five civilian casualties that month and in October were in London.

In those first few infernal weeks new sights and sounds – once unthinkable – were to become familiar. There was never a massed daylight attack quite on the scale of that Saturday – an assault by the largest air fleet then ever assembled – so many of the features of the Blitz were to be experienced and heightened by night. There were the ochre and crimson skies of the smouldering metropolis, the yellow sparkle of ack-ack fire, the pale jade beams of the searchlights. There was the clatter, too, of incendiaries falling on roof tiles, magnesium canisters the size and shape of a beer bottle that burst with a white flash and began to burn at a temperature high enough to cut steel. Dogs set up a plaintive howling mimicking the sirens, while the shrill din of burglar alarms muffled the greedy licking of the flames. By the last day of September, the London Region Brigade had attended ten thousand fires.

In the morning, amid the grit of brick dust, there was the pungent odour of high explosive, a chypre of hexogen and ammonal, punctuated by the hissing stink of domestic gas. From the Thames came breezes reeking of the broken sewers that polluted it, blended with chemicals that failed to stifle the stench. And then there were the corpses. In the middle of the street were the remains of a baby, blown clean through a window. It had burst on hitting the pavement. The bodies identified by grieving relatives at the mortuary were composed beneath their shrouds of assorted limbs kept in a basket for that purpose, so that people would think their loved ones were intact and not merely a grim jigsaw puzzle. One woman found that one of her chief anxieties was 'to be sure to have one's handbag near one' in the event of a direct hit. It contained her identity card. 'My body shall NOT go in one of those ghastly cardboard boxes in which civilians unclaimed are dumped in.'

Then there was the struggle to get to work, through a tangle of electricity cables and telephone lines, over which streamed water from broken conduits. Buses went the long way round. The transport situation was made worse by UXBs. In the first few days three stations had to be closed, and within a few weeks much of central and southern London was without trams. A boat service was laid on instead.

The first (17) to fall on the capital landed in Smith Square, Westminster. There it settled beneath large gas mains and the principal electricity feed to Whitehall. Fortunately it failed to rupture the six-thousand-volt high-tension cable. When its twin blew up in the button store of the Corps of Commissionaires,

Stuart Archer arrives at Buckingham Palace with his mother in 1941 to receive the George Cross. By then, he had made safe two hundred unexploded bombs.

(Stuart Archer)

John Hudson in 1943, holding a flask of liquid oxygen as he prepares for the first time to counter the Y fuze, which had been designed specifically to kill bomb disposers.

(Dick Hudson)

The Earl of Suffolk, with his customary cigarette holder, and his assistant Fred Hards pose in 1940 with a parachute mine at the experimental research site in Richmond Park. *(Crown Copyright)*

Robert Davies and his section in 1940, removing a UXB at the German Hospital, Hackney. The presence of a photographer, and Davies's pose, suggest that the fuze is not live. *(Getty)*

Ludgate Hill cordoned off in September 1940 while Davies and his squad search for a UXB that had landed in front of St Paul's Cathedral and threatened to destroy it. *(Getty)*

George Wyllie, who finally uncovered the UXB, wearing the George Cross that both he and Davies received for their actions at St Paul's. *(Evening Star)*

Workmen prepare to fill in the site of the excavation for the bomb, which after three days of digging was found twenty-six feet down. *(IWM)*

The glint of the Thames helped the Luftwaffe to find its targets during the Blitz. The docks at the Isle of Dogs were among the most heavily bombed areas of London. *(Topfoto)*

More than two million houses across Britain were damaged by German bombs in 1940 and 1941, and more than forty thousand people killed by them. *(Getty)*

A 'Blitzed out' family pushes in a handcart the possessions that they have salvaged. One in six Londoners was made homeless at some point in the war, often by UXBs. *(Getty)*

A bomb that fell on Bank Underground station in January 1941 killed 111 people and created a crater so large that Royal Engineers had to span it with a temporary bridge. *(Getty)*

The ruins of Swansea after three consecutive nights of bombing in February 1941. Seven men from a bomb disposal section were killed by one of the UXBs that had fallen. *(Alamy)*

Sheffield was attacked in December 1940. The façade of the C&A store (centre) was ripped open, and more than seventy people died in the wreckage (left) of Marples Hotel. *(Alamy)*

Herbert Hunt stands beside a 1800-kilogram bomb, the largest in the Luftwaffe's arsenal and nicknamed 'Satan'. Hunt had many narrow escapes while inspecting UXBs with ticking fuzes. *(Topfoto)*

Edward Talbot (right) at Neath in 1940, supervising an excavation for a suspected unexploded bomb. He was among the first bomb disposal officers to win a gallantry medal. *(Royal Engineers Museum, Library and Archive)*

A cheer goes up as a twelve-foot-long unexploded Satan, complete with tail fins, is extracted from forty feet down more than four years after falling on Croydon in 1941. *(IWM)*

This view of the Croydon's fuze boss shows the date when it was manufactured. Other markings indicated a fuze's purpose, such as delayed action. *(IWM)*

A Type 17 electric time fuze and, below it, the Zus 40 booby trap that would explode the bomb if an attempt was made to extract the fuze from its pocket. *(Courtesy of Steve Venus)*

pedestrians in the Strand were showered with what at first they thought were gold coins.

Returning home from rescue duty in Millwall, Bill Regan found his street cordoned off. An unexploded bomb had fallen opposite his house, where his redoubtable wife Vi was tidying up shards of plaster. 'She showed me the hole where the UXB had gone and wasn't concerned about it,' he marvelled. 'Warden Herbie Martin had tried to persuade her, and her mum and dad, to take shelter in Saunders Ness School, but they declined. Vi said she wasn't afraid of dying provided she didn't have a lingering death. Surprisingly, this took away the worry I had at the back of my mind while on duty.'

It was rapidly apparent that the BD presence in London was far too small to cope with the demands that now overwhelmed it. Major-General Taylor, the new Director of Bomb Disposal, had been in his post for less than a week before Goering struck on 7 September. The only BD Company in the capital, No. 5, had formed at Chelsea Barracks on 1 September. Just twelve sections – ninety men either side of the Thames – protected the entire city. Within three weeks it would be throttled by almost a thousand UXBs.

Even after the first raids, it was clear that the threat these would pose was acute. On 11 September a report by the MOHS on the effects of the bombing on the capital's railway network, which had been crippled, ascribed 'most of the trouble [to] . . . delayed action or unexploded bombs, which paralyse movement until touched off or dug out'. An additional problem was that, with the railway tracks out of action, rolling stock was building up in the marshalling yards and presenting the Luftwaffe with another prime target.

In his history of the war, Churchill reminisced about this 'new and damaging form of attack': 'Large numbers of delayed-action bombs were now widely and plentifully cast upon us and became an awkward problem. Long stretches of railway line, important junctions, the approaches to vital factories, airfields, main thoroughfares had scores of times to be blocked off and denied to us in our need.' The time fuze was, he considered, 'a most effective agent in warfare, on account of the prolonged uncertainty it causes'.

Minuting Anthony Eden at the War Office on the MOHS report two days later, he urged him to 'expand as rapidly as possible General King's organisation. I do not think he has planned it on large enough lines to cope with this nuisance, which may soon wear a graver aspect'. Eden was able to reply that the day before the decision had been taken to vastly increase the total of BD sections, doubling both the strength of each (to thirty men) and their number (to 440). Such a quadrupling would raise the sappers employed to ten thousand, but it would take time to recruit and train them, so eleven general construction and quarrying companies were being pressed into service to plug the gap. A dozen other sections had also headed to London to reinforce 5BD, now engaged in what Churchill called 'a task of the utmost peril'.

One important practical step had been taken the day after the first attack. This was the adoption of the system of prioritising UXBs. Category A would be for bombs that it was essential to dispose of immediately, regardless of risk to BD personnel. They would be sub-divided into those that could be exploded in situ, and those that could not. Categories B to D covered all others, from those that needed to be dealt with rapidly to those which could be dealt with when it was convenient to do so, for example if lying in open countryside.

Unless a bomb was classified as Category A, it would be left for ninety-six hours in case it proved to be a (17) delay. Many UXBs would however need urgent attention and pressure was mounting for the scientists to find a means of counteracting those armed with the Zus. Freddy provided a measure of safety, in that the officer did not have to remove the fuze in person, but it was not infallible. Too often its spindly frame jammed at just the wrong moment, with the fuze only partially extracted.

'It wasn't a very popular piece of apparatus,' recalled William

Wells. 'You never knew whether something unusual was going on, and what you might find when you returned to the bomb: perhaps a damaged and dangerous fuze or a mechanical booby trap, needing only a slight disturbance to set it off.' Cursing, the officer would now have a nervous few minutes trying to complete the job himself.

By contrast, the steam steriliser seemed about to offer a more comprehensive solution. As Archer had done at Llandarcy, it would seek to minimise any damage that the fuze might do by stripping it of its punch – the explosive. The only drawback was that the melting process took time and needed sappers to set it up, and all the while the bomb's clock would be running. As yet, there was no way to stop that. London, however, was offering plenty of opportunities to put the remodelled equipment through its paces.

On the Thursday morning of the first week of attacks, Captain Kennedy was told that in the night a 250-kilogram UXB had landed outside the Ford car showroom in Regent Street, near the Café Royal. He asked for a steriliser squad to meet him there, and arrived at twenty to two. So did Merriman, and Gough himself. The bomb had bounced off two buildings and skidded along the tarmac before coming to rest. The street was cordoned off with rope and the yellow signs denoting a UXB that were soon to become familiar, and Kennedy examined the device. It had two fuzes, both of them (17)s, and by using a short steel rod as a stethoscope he could hear that at least one was ticking, though he could not tell which.

Old soldiers both, Merriman and Gough were not to be denied the chance to see their enemy at close quarters. Courageously, if rather foolishly given their importance to BD, they too went to inspect the fuzes. 'I shall not easily forget my feelings,' Gough said years later, 'and Merriman's, when he and I walked out to that bomb, lay down, put our ears to the cold hard case and heard the tick, *and knew we could not stop it*. All we could do was to get as much explosive out as possible before it went up.'

A wall of sandbags was erected and the boiler prepared behind it. By half-past three the steam-powered drill had been fitted to the bomb and had begun to bite into it. After ten minutes it stopped, indicating that it had penetrated the casing, and the steam-injector nozzle started automatically. An hour later it was obvious that it was not working properly, as all that was coming out of the bomb was steam.

Kennedy again approached the device and saw that the cutter had in fact failed to make a hole. He altered its position and tried again, but without success. In desperation, he rolled the bomb over to see if for some reason the drill would work better on the other side, and so it did. But then the switch for the injector broke. Kennedy managed to force it, only for the pump to cut out. This was got going with some hefty blows of a hammer. All the while, the clock was ticking.

By five to six the steriliser was at last beginning to work well, and a thin trickle of emulsified TNT started to drain from the bomb. At six o'clock precisely it blew up. Half of the plate glass windows in Regent Street were splintered by the blast, but though the BD party behind the sandbags were shaken by it, the only injury was a slight cut to a sapper. It had been a lucky and a useful afternoon's work, yielding priceless information about the steriliser's shortcomings. The hard truth was that only by using it in real operations could the apparatus be properly evaluated, and indeed the next day Kennedy was testing it again, in Covent Garden. This time he and his men got out half an hour before the bomb exploded, the machine having played up once more.

An air of unreality hung over London during those earliest days of the Blitz. It was as if the population could not believe that the war had finally caught up with them. Some quickly appreciated that it heralded a new age, one where destruction came suddenly to rich and poor alike and where many of the old ways of deference and obligation would be swept away. Others seemed reluctant to let them go. After the bomb had fallen on Regent Street at three

in the morning, the buildings around Piccadilly Circus were evacuated. Outside the cordon, a Guards officer in a dressing gown
mentioned to a warden that he had had to leave in a suitcase in his
hotel all the possessions he had, having lost everything else in the
retreat from Norway. Without hesitating, the warden ducked
under the rope, walked back to the hotel, retrieved the case, and
presented it to the officer. It was the decency shown amid the
dangers that made them bearable, and gestures like this were to
become commonplace in the coming months.

'At such moments,' wrote one author who experienced the
Blitz, 'having saved a life, and seeing the plight of the rescued, it
seemed that at all costs life should be made worth living, as well
as merely preserved.'

9

SAVING ST PAUL'S

It was the greatest feat of courage of the war, proclaimed the headlines, 'A story that must win a man a VC.' To a nation rocked by the onset of the Blitz, the saving of St Paul's from destruction by a UXB symbolised Britain's defiance of the bombers. For BD, the publicity that followed brought their work out of the shadows and established their status as the people's protectors. Yet at the centre of all this drama was a most unlikely hero.

Bob Davies was almost forty, but looked older. He was of above-average height, but he was thickset and seemed shorter. Only something about his grey-blue eyes would have singled him out in a crowd of other stocky, middle-aged men: there was a jauntiness to them, an optimist's charm. Like his father John, a blacksmith, he had been born in Cornwall and, like him, as well he had found its confines too narrow. John had sought his fortune amid the rugged landscape of southern Africa, witnessing the start of the Boer War before returning to his native Penzance. Young Bob had been schooled there, but by the time he was fifteen he was restless and eager for more adventure than smithing or the life of a trawlerman promised.

In 1916 he sailed for Canada, but almost as soon as he arrived he tried to enlist in the army it was sending to the Western Front. Although he lied about his date of birth he was still held back on account of his youth. It was not until 1918 that he got his wish, and then it proved something of a disappointment. Instead of winning glory in France he found himself in Siberia – in January – as Canada briefly lent support to the White Russians' fight against the Reds. Davies promptly came down with jaundice, and after several months in hospital in Vladivostok was shipped back across the Bering Strait.

By the mid-twenties he had acquired an engineering qualification, a wife – Isabella, originally from Scotland – four children and some exotic stamps in his passport. His travels took him to Japan and Nigeria, and eventually to a construction job with the British West Africa Public Works Department and another bout of ill-health, this time malaria. He was invalided back to the West Country, and by 1939 was living in Plymouth.

His service there as a special constable, his knowledge of engineering and his prior military experience (perhaps not gone into too deeply) all recommended him for an emergency wartime commission in the RE. In June 1940, he was sent on one of the early courses at Manby, and by July he was in command of two of the new and autonomous BD sections, No.s 16 and 17. These were based in the City of London at Bunhill Row, where the Honourable Artillery Company had its headquarters. Aside from drills and training, there was little for the section to do for the next few months. When it was decided that the capital would be better defended if its sections were amalgamated into the first BD Company, No. 5 – given for simplicity the same number as the Civil Defence Region – Davies became third-in-command of the city's BD resources. He was also directly in charge of a third of its strength north of the Thames, although like his brother officers he had almost no actual experience of bomb disposal.

The arrival of the Luftwaffe a week later was the proverbial baptism by fire, and the next few days and nights passed in a blur of feverish activity for all the sections. The fringes of the City were among the areas hit by the raids and Davies was bombarded with multiple reports of UXBs. In between taking shelter themselves, his two squads dealt with their first bombs. Then, on the morning of Thursday 12th, at about the same time as Kennedy was being told of the UXB in Regent Street, Davies received an urgent call at Bunhill Row. A bomb had been heard to fall at about half past two in the morning, close to the front of St Paul's Cathedral. Since it had not yet gone off there was a suspicion that it might be a time fuze. This was a Category A – it must be got out before it exploded, whatever the cost.

Then, as Davies hurried to get a working party together, there was a flash and a blast much closer to hand than St Paul's. A fuze pocket that had been removed from a bomb and taken to the section office for inspection had erupted there, starting a fire and seriously injuring a corporal and a sapper. The flames were soon put out, but all of 16/17 Sections' equipment and records had been destroyed, together with kit and clothing belonging to Davies and another officer. There was no time to do anything about it now, but it was a salutary reminder of the dangers and unpredictability of fuzes.

When Davies arrived at St Paul's he saw that a wide space in front of the steps had already been cordoned off. Judging by the size of the hole in Dean's Yard, the bomb was evidently large and it had narrowly missed the south-west tower before burying itself a few yards from the cathedral foundations.

The police had cleared the area and traffic was barred from Ludgate Hill. Aside from the fire watch, only the Dean, Walter Matthews, had been able to get past the guard. He was due to give a talk for the BBC in a series entitled *Lift up Your Hearts*. 'For some reason,' he wrote afterwards, 'I got it into my head that I could not possibly give the talk without shaving, and I succeeded in evading the police, getting into the Deanery and performing the operation, listening all the time for the roar which would probably announce the imminent collapse of the house. Looking back, this seems to me a very silly thing to have done.' Otherwise the church stood silent; for the first time in centuries the daily services went unperformed.

A scan of the scene revealed how potentially damaging the bomb could be. Not only did the angle of its impact suggest that it must have burrowed towards the magnificent West Front of Wren's masterpiece, but it had also landed atop the entire trunk telephone network to the North of England. If it was a (17), then it had already been ticking for twelve hours out of an accepted maximum of about eighty.

Time was all-important, and the six sappers with Davies jumped into the crater made by the bomb. They began to chip away at the paving above a jumble of wire and cable that was now visible – then in a matter of seconds they went down like skittles, unconscious. In passing through the ground the UXB had fractured a six-inch gas main, and the soldiers had been overcome by carbon monoxide fumes in the hole.

Nearby was a party of men from the Gas Light and Coke Company who were already working on breaks in the pipe. They found some of the sappers about to be given milk, which was popularly believed to be the right treatment for those who had inhaled coal gas, such as would-be suicides. This they knocked away, knowing instead that what was needed was artificial respiration. Even after this had been tried, however, the soldiers had to be taken to hospital, although none was detained for long.

Meanwhile, the gas workers began to cut off the supply. By ignoring the usual precautions, in a few hours they were able to do what would normally have taken two days. Even so, a fire broke out and they had to get the brigade to temporarily flood the main with water. Then evening came and the raids resumed.

The best part of a day had passed before the men from 16/17 were able to start digging. Through the night they laboured methodically, through gravel, dirt, sand and mud. Soon after dawn on Friday three of them were burned when a spark accidentally ignited another gas leak. Everyone threw himself to the ground – all except Davies, who remained standing, seemingly unperturbed, on the rim of the crater. After being treated nearby at Bart's Hospital, the trio were able to return to duty. On they worked, now through black clay polluted with fumes from the main, the clink of their picks the only sound on the hill.

When they got down about twelve feet, the bomb was spotted. The friction generated by its passage through the clay had given it a high polish, encouraging it to slip ever deeper under its own weight. It also made it hard to grasp, and when an attempt was made to pull it out there was a near disaster: it squirmed from the hawser like a huge greased pig, and disappeared still further into the mire. There was nothing for it but to start digging again.

Down and down they went, hour after hour, encouraged by Davies. Occasionally they made a dash for cover as the whistle of falling bombs was heard. Saturday came and went, and with it the first news report of a UXB at the cathedral. In case they needed reminding of the danger they were in, an article in the *Daily Mail* noted that 'These most gallant – yet most matter-of-fact – men of the RE are many a time running a race with Death.'

By Sunday morning, almost eighty hours had elapsed since the bomb had fallen. Away to the west, raiders could be seen dropping bombs on central London. Then from the pit came a shout, and the sharp ring of metal on metal. At twenty-seven and a half feet underground, Sapper Wyllie had uncovered the monster's lair.

And it was a monster, 'a vast hog' eight feet long and weighing a ton, lying at the end of a tunnel no wider than a man's shoulders. A half-inch wire rope was eased around it from pulleys attached to a truck, but its ponderous bulk was held fast by the clay. Finally the strain from the winch was so great that the line snapped. When the operation was repeated the same thing happened again.

It was now midday and, in almost frantic haste, another lorry was paired in tandem with the first. This time, the thousand-kilogram bomb rose slowly out of the crater like a diver magically drawn from the water, before being lashed tightly to a cradle.

Preceded by police motorcycles, the truck sped as fast as it could through the streets of East London until it reached Hackney Marshes. There, the bomb was detonated by Davies. It blasted a hole a hundred feet wide in the ground and shook the plaster from houses far away. The danger had been faced and bested, and rising above the smoke and flame, St Paul's would come to signify resolve and survival.

The fillip that the news of its miraculous salvation gave to Britain's battered morale was just the story that the press needed. 'London is wild with praise for them,' wrote one woman, having seen the flood of articles about Davies, his men and the 'biggest bomb ever dropped' on the capital. The section became celebrities overnight.

There was Serjeant George Wardrope, a small and alert figure who had spent his birthday quarrying for the bomb. 'You couldn't have a better lot for such risky work,' he said by way of tribute to the others. Sapper E. J. Norman, a married man of thirty-six who had helped in the final excavation, told readers that they had never been frightened: 'Scared? Well, we just dig it up as if digging the garden and we take no more notice of a bomb than if it were a worm.' Nevertheless, for the papers they were still a 'suicide squad' who had 'saved the cathedral from almost certain destruction by a gigantic German time-bomb'. So grateful had been one of the canons of St Paul's that he had ventured into a public house to buy them all a pint.

George Wyllie, who had been the only man at the bottom when the UXB was found, was a dark-haired Scot from Kilmarnock. Though the journalists managed to extract some colourful quotes from him – 'I don't mind being gassed. I don't mind being blown up. But I don't *blue pencil* well like being electrocuted!' – he was by nature taciturn. A high proportion of those drafted into BD were Scots, and his background was typical of many of them.

Thirty-one years of age, he had only been a few months old when his coalminer father had died, leaving a widow, eight children and few savings. At twelve, George had gone out to work.

He had had a series of jobs, including as a delivery man for the
Co-op and building electricity pylons for the new National Grid,
when at eighteen he suffered a serious injury in a football match.
His kneecap was torn open in a tackle, and within two weeks it
had swelled to three times its size. The surgeons diagnosed blood
poisoning and wanted to amputate his leg. He had endured
twenty operations and a year in hospital before being able to
hobble home on both feet.

When the Depression came to Ayrshire Wyllie had moved
south in search of work. Until falling out with the foreman he had
found it at the Avro aircraft factory in Manchester. Then he had
made ends meet as a pipe fitter until the war arrived. One look at
his broad chest and his experience in the Territorials had secured
him a berth in the RE in August 1940, but after a couple of
weeks at Chatham Barracks he was desperate to get out. Just nine
days before his heroics at St Paul's, he was assigned to bomb disposal duties. 'When I found out what the job was,' he told an
interviewer, 'I had no regrets. If it goes up, then I go up with it.'

There was more official recognition of his deed, too. In a press
release the Ministry of Home Security stated that 'only the
courage and tenacity of the officer, his NCOs and men prevented
St Paul's from being levelled to the ground'. Eden, the Secretary
of State for War, praised the 'devotion to duty' of all BD units.
The Times, meanwhile, saluted in a leader 'the outstanding deed of
heroism so far recorded in the capital', noting that the 'new turn
of the war . . . has placed every inhabitant of London in the line
of battle'. There was a public service of thanksgiving in the
Cathedral.

Inevitably, however, much of the attention focused on the section's officer, Davies. A newsreel was to be shot about him and
shown in America, while one paper had tracked down his in-laws
in Canada. 'It is just like Bob,' they said. 'If he was going to do
something, nothing could stop him. He always was a dare-devil,
and clever.' When it was revealed that his son Jack was also in the
Army – and had been wounded at Dunkirk – it seemed as if the

whole family was made of the right stuff. 'It was just in his line,' reckoned his wife when asked about his exploit. 'He is very handy about the house.'

His 'cold-blooded disregard of supreme danger' had thrilled the country, but though it showed them that the bombs could be beaten, it did not slow the rate at which they continued to fall. There would be much more work for BD officers in the weeks to come, and not all of them would have Bob Davies's luck.

10

RUN, RABBIT, RUN

By late September, London was beginning to adapt to the Blitz. The savaging on the afternoon of the 15th of a large formation bound for the docks showed the Luftwaffe that they were vulnerable by day, so from then on significant raids took place only at night. In any case, the arrival of the autumn rains meant that good daytime visibility was less probable. The moan of the siren was heard more rarely before dark; it was dawn that brought the scent of powdered brick and the crackle of glass splintering like ice underfoot.

For some, the blackout cloaked the city in a different beauty, one illuminated by the yellow glare of a burning main. There were other novelties too. Buses in unfamiliar green or purple liveries appeared, drafted in from the provinces to replace those destroyed in their depots. And the telephone system had become strangely unreliable: a secret policy ensured that private calls were cut off or not put through once bombing began, so as to leave the lines free for the emergency services.

After a confused and often ineffectual start, these services had begun to perform better and to coordinate themselves more intelligently. Unspectacular tasks turned out to be of great practical importance. None was more vital than ascertaining how many people had been in a building before it collapsed, potentially saving much useless and hazardous searching amid debris. Foot and bicycle messengers proved more reliable than fragile telephone wires for carrying news to ARP Control Centres, and some wardens began to learn how to distinguish false reports of UXBs from those that were genuine. In time, with the scientists and the Forces, the emergency services were to become the third apex of the warning triangle that stood between the civilian population and the bombs.

Like the post, the milk and the newspapers, Londoners generally carried on much as before. Many of those who could, of course, especially the better-off, had already left the city. By the end of November 1940 the population of the twenty-eight central boroughs would drop by a quarter, while by Christmas five in six children (and many of their mothers) had been evacuated.

Far more of those who stayed had lost their homes than had been predicted: some fifty thousand houses per week were damaged in September. The impact of seeing one's possessions destroyed – a lifetime of memories eradicated – could be almost as devastating for many people as losing a family member. Gwladys Cox and her husband Ralph lived in a mansion block in West Hampstead. On 1 October they took shelter from a raid, and returned the next morning to find that their home had been gutted by an incendiary bomb: 'Ralph's beautiful antique desk which, besides our marriage certificate and Fire Insurance Policy, contained all the letters I had written to him . . . not even the ashes remained.

'His hundreds of books, his chief hobby, collected during a lifetime, were congealed black masses of cinders; hundreds of gramophone records had vanished into thin air.' Disconsolately, they wandered from room to room, their feet squelching in the saturated carpets. 'In the drawing room, the roof had fallen in and what remained of the sodden furniture was covered with a shining layer of molten lead from the burning roof; the carpet was inches deep in a wet mixture of ceiling plaster and burnt rafters.'

As if being violated by the Luftwaffe was not bad enough, they were later to make another saddening discovery. 'I found that my silver cigarette lighter had disappeared from its drawer. Under my bed, my trinket box was lying open, its contents scattered over the wet carpet. It had been taken out of the dressing-table drawer, which had been forced. Ralph's room had been ransacked and most of his underclothing, as well as a gold watch, taken. All this the work of looters.'

When entire neighbourhoods of London were forced out by UXBs, the few rest centres soon overflowed. Yet the numbers killed – 5500 in the space of three weeks – had been far less than the Government had feared. This was all the more surprising given how many people had not accustomed themselves to seeking shelter at night.

If there is a myth of the Blitz, it is the enduring notion that the average Londoner survived it by huddling in a public shelter with his neighbours, an experience that bred a spirit of togetherness. In fact the communal shelters, of which five thousand had been built by local councils, were unpopular almost from the start. Not having been designed for lengthy occupation, they had no lavatories and poor ventilation. Moreover, they were quickly and correctly perceived as unsafe. A shortage of cement and an absence of scruples had meant that many had been constructed in haste by contractors willing to use just lime and sand as mortar. As a result, should a bomb fall too close, the walls often collapsed and the reinforced roof crushed those inside.

In the first few weeks of bombing, when the shelters were widely used, they soon became crowded. In addition, there was also much resentment that many of the municipal ones were only open during the day and, for reasons best known to officialdom, closed at weekends. People began to seek alternatives. The basements of modern steel-framed buildings such as apartment blocks and department stores were safest. Upwards of fifteen thousand slept in a warehouse under the railway arches at Stepney. Above all, Londoners began to bed down on platforms of the Underground, favoured largely because of their warmth and their depth, which rendered the raids inaudible.

Initially there were attempts to prevent the stations being spontaneously turned into dormitories on the grounds that having to step over slumbering forms inconvenienced commuters. Plans were made to ban the 'Tube squatters', but by the end of September more than 175,000 people were sleeping there and opposition to the evening migrations began to seem as crass as it was. Official sanction was given once a young engineer at the MOHS, Derman Christopherson, had used a map of the Underground and statistics from BD about missile penetration to prescribe which stations could safely be used as shelters.

The American war correspondent Ernie Pyle visited Liverpool Street station to see its night-time inhabitants and their troglodytic life. Many came from the working-class areas to the east, and stood to lose what little they had in the bombing. Their plight touched him deeply: 'Many of these people were old – wretched and worn old people, people who had never known many of the good things in life and who were now winding up their days on earth in desperate discomfort . . . There were children too, some asleep and some playing. There were youngsters in groups, talking and laughing and even singing. There were smart-alecks and there were quiet ones. There were hard-working people of middle age who had to rise at five to go to work.

'Some people sat knitting or playing cards or talking. But mostly they just sat. And though it was only eight o'clock, many of the old people were already asleep. It was the old people who seemed so tragic. Think of yourself at seventy or eighty, full of pain and of the dim memories of a lifetime that has probably all been bleak. And then think of yourself now, travelling at dusk every night to a subway station, wrapping your ragged overcoat about your old shoulders and sitting on a wooden bench with your back against a curved wall. Sitting there all night, in nodding and fitful sleep. Think of that as your destiny – every night, every night from now on.'

Nonetheless, only about one in ten Central Londoners used some kind of public shelter. Three times that number preferred either a private refuge, such as a shop basement, or one of the 2.5 million Anderson shelters that had been given away by the government to lower-income households and erected in back gardens. Even so, perhaps as many as two-thirds of all Londoners stayed in bed at home when the siren sounded, or reported for duty with the ARP and emergency services.

'It is true and comforting,' wrote the author John Strachey, who spent the Blitz as a warden in then bohemian Chelsea, 'that the reality is never as bad as the fantasy: never as terrifying, never as shattering. That is why the raids, when they came, were such a genuine relief to so many people.' There was a collective shrug of the shoulders by Londoners, a mixture of resignation and resilience, a recognition that they could cope.

As many noted, that communal necessity, and the dangers that provoked it, dissolved social boundaries. The writer Constantine FitzGibbon observed that the middle-class Englishman 'does not like his neighbours to know what he is doing: he even prefers his neighbours not to know who he is . . . A mumbled "good morning" on the doorstep, a non-committal banality about the weather in the lift, these are considered perfectly adequate acknowledgement between people who have lived, for years on end, within feet of one another.'

FitzGibbon found that the bombing ended such reticence. For one thing, everyone had a common experience of danger to share. 'Everyone had his story and if other people's were obviously less interesting than one's own, it was at least worth listening politely while awaiting one's own chance to tell. Contact was established.'

His wife Theodora believed that, as a result of this breaking down of reserve, a new spirit of friendliness was born in the Blitz. 'People were concerned with helping their neighbours: there was a joke or a laugh to keep their spirits up, and a sharing of scarce commodities. The last pinch of tea, or a bottle of whisky, was offered by people one had never seen before and might never see again.

'Everybody was in love with life and living. In apartment houses, the owner of the basement rooms expected to share his or her bedroom with perhaps five or six persons of mixed sexes. Once you have lain on a bed, even platonically, with someone for several months, it is impossible to ignore them afterwards.'

But even basements were not invulnerable. For the unlucky, there would be a moment of panic as the walls started to collapse on them. Then there would be the sound of thud upon thud, as tons of falling masonry raised the pile still further, entombing together the dead and still living. Bill Regan's heavy rescue squad had to deal with the aftermath of such tragedies, and on one occasion he and his friend George were called to an Auxiliary Fire Service building that had been hit some days earlier.

'George called me to help him with a doormat he had found but could not pull clear. It was black and of a thick curly texture, so I fished around for a while, loosening the rubble. Then George came back with a length of iron to prise it out. I told him it was a bloke, and I knew who he was – Warden Herbie Martin.' It was Martin who had earlier tried to get Vi Regan to take shelter from the UXB outside her house.

Their workmates had made another discovery. When the bomb had landed, two young firewomen had been asleep in the dormitory. They had been buried instantly beneath the rubble and so densely packed was it around them that rigor mortis had not set in and they seemed still to be asleep. 'I looked around at the other men,' wrote Regan, 'and most of them looked shocked and a bit sick; we had usually found bodies mutilated, and they were usually lifted out by hands and feet and quickly got away. Major Brown sees one man being sick, so he fishes out a bottle of rum to be handed around.

'By now I was feeling a little bit angry at the prospect of these two girls being lugged by their arms and legs.' They were dressed just in their knickers and slips. 'I could not let them be handled like the usual corpses. I know I would have belted the first one who handled them with disrespect, but nobody makes a move to shift them and are just standing there, gawping. I looked up at George, and I just said "Stretcher – blanket." Then I put my right arm under her shoulders with her head resting against me, and the left arm under her knees, and so carried her up. I laid her on the stretcher: "You'll be comfortable now, my dear." I did exactly the same with the other one . . . I cooled down a bit after I had smoked a cigarette. I wonder why I had been so angry.'

The experience of being trapped by such an explosion was for most people traumatic. Dorothy Rothwell was a land girl in Kent when one evening she heard gunfire. She ran to the window to see what was happening, and then her world turned black. A fifty-kilogram bomb had hit the house where she was lodging. It had gone right through her bedroom and exploded in that below. Two women had been killed, while she was trapped under the debris.

'I was unconscious at first but then I came to,' she remembered. 'I was upside down with my watch in my right hand, and I tightened my grasp on it. But apart from that I couldn't move at all. I was completely buried. I prayed to God that they'd find me. After a couple of hours, they dug down and they got my right leg out up to my knee and I felt a morphine injection go in. They kept digging until I just smelled fresh air.'

Dorothy had shrapnel wounds to her shoulder and neck, and a piece of brick was embedded in her chest. Her left arm had been pierced by a falling beam and they had to saw through the wood before she could be freed. She had also broken an ankle. 'But my feeling was just relief.' It was more than a year before she was well enough to return to work, and even then, she said, 'I was very nervous when an air raid was on.'

The Blitz was a particular strain on those who were separated by duty from spouses or family still in London. Men felt unable to comfort their wives during air raids, and news of their safety was often slow in reaching them afterwards. Reg Dance was a young private in the Rifle Brigade. He had decided to marry his girl-friend Maisie after being evacuated from Dunkirk. Following their wedding at the garrison church at Woolwich in January 1941, he had returned to his unit's billet in Essex. Two weeks later he was told by an officer that Maisie had been severely injured by an oil bomb that had landed on the electronics factory where she worked.

When he arrived at her bedside at the hospital, Dance could barely recognise his wife. She had been proud of her auburn hair, which was now matted with oil. Her face was badly bruised and her cheeks were still pitted with fragments of stone. 'She was crying with pain,' he recalled later, 'because her back was shattered and her arm was shattered.'

Her wounds soon became infected, but Dance had to return to the regiment. Three days later he received a telegram calling him back home at once. 'I went straight to the hospital and I could see she was terribly ill. She had developed tetanus. Her whole body was rigid. Her face was locked. I don't even think she knew I was there . . .

'I make no bones about it, I dropped to my knees and I prayed to God that she would not suffer any more. Then I walked out to get a nurse, and as I walked back in, she died.'

As the human cost mounted, scenes such as this took place all over the capital. Fathers were forced to identify the bodies of their young children. Husbands were left without wives. People could just about cope with their own fears and the terror that the bombs brought. What they found far harder to bear was the death of someone they loved, and the absence of a reason for it.

Andrew Butler was an architect who assessed damage to houses for the council in Chelsea, and he recorded one conversation that he had had with a man who had lost his family. 'I admit I found the recital almost overwhelming,' he wrote, 'and would have gone outside and howled or broken things and behaved like a child but for the quality of severe grandeur in the man's bearing.

'He had got, it seemed, beyond mere hatred of the enemy and had touched a kind of frozen detachment regarding himself; but he spoke very passionately in short staccato phrases when he said,

"This isn't war at all. It's plain murder of ordinary folk. Why should they kill Mary? – and the girl? They didn't start the thing. Nothing to do with it. And the parson said she died for her country. I'm sick of all that talk. What's the good of it?" . . .

'Then he stopped and I saw he was looking at an old photograph of his wife which was hanging by the fire. He gave a short gasp and sat quite still, just staring, in the fading daylight. Only an occasional drip from the sink's tap broke the silence in the room. A room like a lot of others, I expect, all over Europe, with a plain man or woman in it enduring the maximum sorrow, and wondering why.'

The capital might absorb punishment well enough, but there was little that it could do about it being dished out in the first place. When the aerial assault began London's defences were found wanting. The RAF could stop the daytime raids, but with radar sets still in their infancy interception of the bomber streams at night was almost impossible. Nor was any greater protection afforded by the city's anti-aircraft batteries.

During the first two or three attacks Londoners had been puzzled at the spasmodic nature of the flak directed at the aeroplanes overhead. In fact, the technology for estimating a target's elevation and bearing, the Fixed Azimuth system, was cumbersome and outdated. Its computations were based on aircraft noise relayed by sound locators to a central control room, but rested on the assumption that the target would fly on a straight course and at a constant height and speed. The German bombers did none of these, and consequently the chances of the anti-aircraft guns hitting one were minute, and not improved by darkness. From 10 September huge barrages were organised, with every gun told to fire every shell, but this was simply to make a noise that would buck up morale. When, on the night of 15 October, out of 235 bombers two were brought down, for an expenditure of 8326 rounds, it was by some distance Ack-Ack's best shooting to date.

Human nature being what it is, although neither the flak nor the searchlights posed much danger to the German pilots they did make them fly higher. This did not stem, however, the flow of bombs, guided by the flaming beacons that rose nightly from blazing wharves and warehouses. The consequent flood tide of UXBs could not be held back by the dozen sections of 5BD Company alone.

They had done what they could after the earliest raids, notably at Buckingham Palace, which was bombed three times in the first week. 15 BD Section, under Second Lieutenant G. Pringle, worked for much of 9 September to try to excavate a five-hundred-kilogram UXB that had landed on the North-West Terrace. It was thought sensible that the King and Queen spend the night in the country, and at about half-past one in the morning the bomb duly exploded. The blast broke all the glass at the back of the Palace, tore open the walls of the swimming baths and left a crater eighteen feet deep in the lawn. Pieces of the bomb were presented to the King as a souvenir by Pringle. It was to be the first of several visits.

By the middle of the month, with dozens of new UXBs being reported each day, reinforcements began to arrive. 2BD Company, which had formed at Leeds a fortnight earlier, was now given charge of all operations south of the Thames. The existing units there were transferred to its strength, while other newcomers were attached to 5BD, so that both companies were now composed of fourteen sections.

At the same time, a new hierarchy was being established for BD nationwide, which was ultimately to comprise twenty-five companies of twelve sections each – largely those brought into being by the recent quadrupling of the organisation's numbers. In turn, those companies in the south and east of England were marshalled by area into four Groups, approximately the equivalent of a conventional Army regiment. 1 BD Group was in London, with headquarters at Prince's Gate, in South Kensington, and by the end of the year it mustered six companies. Those companies further from the Home Counties, such as No. 8 at Cardiff or No. 11 in Edinburgh, remained independent of any Group.

Another innovation was the creation from the early autumn onwards of volunteer civilian disposal squads. These were formed mainly by workers at large factories and were intended primarily to reduce damage to these. In practice, however, the teams also gave valuable aid within their towns after especially heavy raids, and in time were to be supplemented by auxiliaries from the Home Guard. As they needed training, and priority for that was given to the Forces, their numbers were never great, though not insignificant. About 350 civilians passed through the Army BD School in the first year of the scheme.

Yet even new BD officers were finding that what they were being taught was of little help to them. Cliff Green had recently been commissioned in 716 RE Construction Company, one of those being pressed into BD service. The son of a police sergeant, he had grown up in West London and had then joined George Wimpey, the firm of builders, as a clerk. When war came he was working for Monk & Co. of Warrington, supervising contracts for aerodromes. Like many of his contemporaries, he had thought conflict with Germany inevitable and was already a member of the RE Territorials.

He went to RAF Manby in September believing that he was going on an armaments course. Having spent a week learning a little about bombs, he followed his company to their new billet in Tunbridge Wells. 'We got there on Sunday, and the following morning we took over our sections and checked our stores. Then we were given a list of bombs and a map, and were told to go and deal with them.' Many other officers had equally brisk initiations into BD work.

Rex Ovens, twenty-five, was stationed in several parts of London during the early weeks of the Blitz. Bored with his job as a surveyor in Worcestershire, and unenthusiastic about his marriage, he had sought escape in rugby and the Territorials. He also felt ill-prepared by his BD training, although he did remember some more general advice when caught in the open by an air raid in Acton on his first morning in the capital: 'The best thing to do was to get as close as possible to the ground. With this in mind, when the bombing seemed to be at its most intense and right on top of us, I rapidly got myself into the nearest road gutter.

'While I lay there on my stomach, with the back of my head protected by a steel helmet, I felt a nudge in my back, and when I looked up I saw a diminutive ARP lass with a tray of tea, asking me, in a broad cockney accent, if I was going to stand up to drink it or should she put it on the ground near me? I have to admit that I did not feel much of a hero at that particular moment.'

As new BD officers quickly appreciated, the risks from UXBs were both real and unpredictable. Since the incident at Hook casualties had begun to mount. Sixteen sappers were killed in August, and twenty-nine in September. These last included six members of 5BD Company caught in a blast on the Great West Road, and the CO of 38 BD Section, Second Lieutenant John Hunter, who died when a fuze exploded in his office at Mill Hill Barracks.

Many of the dangers became familiar anxieties. Sometimes the locking ring around the fuze head – three sixteenths of an inch of hardened steel – would not yield and would need to be prised apart with a chisel and mallet. All the while the officer would be hoping that the blows did not start some temperamental mechanism inside the bomb. Fuze extraction could also present problems. William Wells remembered that one friend of his who had pulled out a (17) with a long cord returned to the device to find not just the fuze and clock still easing from the pocket, but also the trigger of a Zus. Thinking quickly, he wedged this in place with a splinter of wood and recited the 'magic words: Quiet, you bastard.'

One of the first jobs that Cliff Green had to deal with was a clutch of UXBs around the Bowater paper mill at Gravesend. A perk of this location was the opportunity, according to a report of the incident, for 'good relations . . . between the men and the female staff' of the factory canteen.

The last of the stick was tracked down to a back garden. The nose of the bomb was visible, but the fuze had been damaged and was unidentifiable. Green managed to chisel apart the ring, but when he removed it most of the fuze head came away in his hand. Using a stethoscope that he had borrowed from a local doctor, he listened to what remained in the pocket: it was a (17) and ticking. Having told the rest of the party to take cover, Green then carefully levered out the fuze with a screwdriver and quickly unscrewed the gaine. The delay mechanism fired just ten minutes later.

Despite the uncertainties, Green regarded his time in BD 'as the happiest that I had in the war. We were convinced we were doing a useful and necessary job . . . We were keen, proud of the Corps, and faced danger with a happy acceptance.'

Of the five officers in Ken Revis's hut on his training course, two were to be killed and another severely injured while in BD. Aged twenty-three, Revis was the son of a civil servant and had been training as an engineer before the war. A day or so after his UXB lectures had finished he was faced with his first live example. It was a five-hundred-kilogram bomb in a garden in Hastings. It was fitted with a simple impact fuze, but this was still daunting enough for a novice.

The first problem was the locking ring, which refused to budge. Perspiring a little, Revis gave it a series of taps with a timbering wedge. Gradually it began to turn, first a half and then a full circle, until it was unscrewed. The fuze was visible in the pocket, and to stop it from falling out before he was ready he slid a folded newspaper down the side.

He was now half-way done. Taking a length of line, he attached this to the boss and fed it through the handle of a spade dug into the lawn. The other houses in the street had been evacuated so, still clutching the cord, Revis retreated to cover and gave it a tug. It was unexpectedly taut. He could hear his heart beating. A bird was singing in a tree. Another pull, and this time the string slackened. Automatically, reflexively, he ducked down, but nothing happened. When he approached the shaft he saw the fuze lying snugly on the newspaper. The worst is over, he thought to himself: if I can do one, then I should be able to do them all.

Most of the new intake of officers found their way to the RE by chance, but it had always been Eric Wakeling's ambition to be in the Engineers. From school at Repton he had sat the Army exams at eighteen. Two months later war had been declared and he thought that he had missed his chance on hearing that the military academies at Sandhurst and Woolwich were to close for the duration. When the new officer cadet training units (OCTU) opened, offering a commission in six months instead of two years, he signed up, and then volunteered for BD after growing tired of his initial posting to a chemical warfare section.

His memories of his early days in BD chime with that of other subalterns: 'There was very little training and very little equipment to start with . . . I think you were well aware of what you were doing and the risks you were taking. But it was a job, and possibly less dangerous than the trenches or the desert or the jungle. I was never shot at and I slept in a bed every night, which is more than a lot of people can say. The dangers were there and your life was more or less in your hands.'

A common rumour at the time was that officers could only expect to survive for three or four months in bomb disposal. It was not a perception borne out by the statistics, but indicative nevertheless of how hazardous the work was thought to be. 'I think you found out quickly how dangerous it was,' considered Wakeling. 'I think somebody once said that life expectancy was about sixteen weeks, because we were dealing with the unknown . . . After 1942 we lost very few officers because we had the right equipment and the right knowledge. But it was a very steep learning curve.'

While officers had at least a general conception of the workings of bombs and fuzes to help them when they began their work, their sappers did not. Most had been called up from civilian life and had only three months' basic military training, much of it marching drill. In William Wells's section, brought down to South London from Scotland in September, there were two shop assistants, a draughtsman, a cinema operator and a couple of labourers. The senior corporal – lance-serjeant was the RE term – was a bus conductor. The serjeant, like many senior NCOs, was a skilled tradesman, in this case a bricklayer. One advantage was that most of the section came from Ayrshire, which quickly bred camaraderie. Wells soon acquired a Scots accent so as to make himself more easily understood by his men.

The future Field Marshal Sir John Chapple spent much of the war as a boy in East Anglia, where his father Charles was an officer with 4BD Company. Like many of the first batch of officers assigned to BD, he was an Army veteran, having served in Mesopotamia and Afghanistan twenty-five years earlier; now fifty-five, he could be spared front-line duties. His sappers were another mixed assortment. Serjeant Kirk had worked for a gas company, while Lance-Serjeant Westlake was a Jew from Brick Lane in East London. Chapple's batman, Macintyre, had been a milkman in Bradford. A reluctant soldier, and an under-nourished one, Macintyre's introduction to the rough humour of the Army had been the recruiting officer's quip to him that if 'he had had a good shit before his medical' he would have been failed for weighing too little.

As many sappers discovered, morbid jokes were the best way to cope with the burdens they were taking on. Aside from the dangers involved, much of the clearing up before and after jobs fell to them, and it was often unpleasant. Among those who witnessed terrible sights was Cecil Brinton, who had joined the RE after Dunkirk. He had no training in bomb disposal, and his introduction to it was an order to go to Clapham Common in London, where he was told to start digging for a suspected bomb.

Often, when BD arrived at the scene little had been touched for fear of triggering an explosion. On one occasion, Brinton's section was called to a house where a UXB had been reported. A crowd of people who had been evacuated from their homes was gathered at the end of the street. One woman appeared especially distraught. Brinton entered the house and defuzed the bomb. Then he went upstairs to see what damage it had caused in falling.

In an upstairs bedroom he found a shambles, and the reason for the woman's agitation became plain. 'This lady had left her baby in a cot in the bedroom. The bomb had gone through the roof right into the cot. It smashed the cot and the baby to pieces. I got a pillowcase and picked up what I could of the baby. I put it in the pillowcase and left it there.

'I told the ARP warden and the policemen what I had done and said I couldn't face the mother. I know what they mean by trauma. It can have a lasting effect.' Sappers might keep their fears for themselves under control, yet they could not but be affected by the damage they saw inflicted on others. Their generation was not always comfortable sharing with others the strong emotions that they felt. Even years later, however, the memories of the brutal and sudden nature of the deaths that they had witnessed remained as raw as the day they had happened.

George Ingram also saw some dreadful sights while in BD. Born and bred in Sheffield, the son of a warehouseman for a builder's merchant, he had left school at fourteen to become an apprentice for a heating firm. He too had joined the Territorials before the war. He had been at work when he heard Chamberlain's broadcast: 'It came as no surprise, but we were all unprepared for it.' While training to go to France with the British Expeditionary Force, he had had an accident while practising building pontoon bridges, and after leaving hospital had been sent instead to 22BD at Chelmsford.

After one raid on Lowestoft, there were so many UXBs that the members of the sections in the area were split up into pairs and told to work independently. A young officer, Lieutenant Ian Hoare, was fortunate to be pulled alive from a shaft that fell in on top of him while he was examining a bomb. The next day, Ingram heard a loud explosion near where he himself was busy with a device. Hoare had been killed, together with an NCO, Geoff Gibbs, probably by a Zus while trying again to defuze the bomb from the day before. 'It blew the corporal over the house,' remembered Ingram, 'across the road, and into a back garden on the other side of the road.'

Gibbs died instantly. Ingram was then detailed to search for Hoare's body. 'Me and this other feller worked on the site from early morning until about half past four at night, but we never found anything, any flesh at all. There was nothing – blown to smithereens.' Hoare's funeral was held at Norwich Cathedral. 'I expect they put sandbags in his coffin.'

Perhaps because they felt that they had less control over their fate than did officers, fewer sappers were inclined to be sanguine about the risks involved. The principal dangers to them came from ignorance and chance. While it was the officer's responsibility to deal with the fuze on his own, they were exposed to it all the while that it was being uncovered. The volume of work, too, was such in those early months that NCOs and sappers sometimes took out fuzes if the officer was busy elsewhere, despite their potentially fatal lack of knowledge about them. When Cecil Brinton's officer lost his life he began to defuze bombs in his stead. Only later did he discover that there was more than one type of fuze. Not being an officer, he had not been sent any of the technical bulletins by BD Headquarters: 'I had no idea what I was doing, and many people got killed doing just that.'

Sheer good fortune spared George Ingram during one of his first excavations for a bomb. 'We started one day and didn't get it finished,' he recalled, 'so we started next day again – it were a case of pick and shovel. When it came to about ten we knocked off for a coffee break, and we were sat back about 150 yards away. Some had gone for a cigarette.

'Suddenly it exploded. Sent showers of big clods of earth that hit me, the blast was real hot as if you had opened a furnace door. When you stand in front of one it sears your face. My ears started to bleed. I didn't know then it would affect me, but it has – sent me deaf.'

Len Jeacocke was another sapper who rode his luck throughout the Blitz. A Londoner, he had born in Brixton in 1915. He had started work at fifteen, and by the outbreak of war was a ledger clerk in a gas lighting firm. Keen to get some military training, in 1938 he had applied for the RAF but found the queue too long. The Navy then offered him a post in the paymaster's office, but this did not thrill him either. His loyalties were at last decided when a major in the TA took him to the pub.

In June 1940 Jeacocke was assigned as an NCO to one of the first BD sections, No. 21, based at a former orphanage in Wanstead in East London. Over the next year he was to work on eighty-two UXBs, from Biggin Hill to Lord Astor's house in Park Lane.

Soon after he began, he and Max Blaney, the section's commander, were caught in a raid while at an RAF barrage balloon site in Silvertown. Terrified, they crouched in a shelter while all around them the docks were pounded by several hundred aircraft. The bombing reached such a pitch that the door of their shelter was blown open; one of the RAF men had to be restrained from running outside after being driven half-mad with fear. That night Jeacocke was so gloomy about his prospects of surviving much longer that he wrote a farewell letter to his mother, thanking her for having looked after him.

So many UXBs had been dropped on East London that there was a 'mad rush' to get them cleared. 'Officially, only officers dealt with fuzes, but they had so many to deal with that the NCOs dealt with them quite a lot,' Jeacocke recalled. The first time that this happened to him was at Wanstead itself, when two of his men trundled up to him with a fifty-kilogram bomb lying in a wheelbarrow.

'Two Welsh ex-miners in the squad had dug down to it in somebody's garden,' he later wrote. 'They had then picked it up, breaking every safety rule in the book, hoisted it out of the hole, put it in the wheelbarrow, and pushed it to an allotment outside the George Hotel. They thought that would be a handier place to deal with it.'

Jeacocke telephoned an officer, only to be told to sort it out himself. 'I approached it, shaking like a leaf. Then, and always afterwards, I had that "here we go" feeling. There was a tremendous sense of loneliness.' The fuze was a (15), but when he tried to pull it out it stuck. Perhaps there was a booby trap underneath? He had a cigarette, and contemplated his options. Then, with both hands trembling, he gave it a tug.

Out it popped. 'I think I actually jumped for joy. Afterwards, you always felt tremendous.' A collection was raised by the grateful local residents: 'One chap put in ten shillings, so we all went to the pub.'

Sappers dealt with their nerves in different ways. Some took to chalking 'Suicide Squad' on the tailgates of their lorries. Jokes and sarcasm were another solution. Rex Ovens could recall an episode when his section had finished with a bomb and were wearily loading its carcase on to a truck. When the householder demanded that they fill in the large hole that they had left in his lawn, one of the sappers replied deadpan: 'Don't worry, boss, the bloke from Kew Gardens will be around later to straighten things out and ask you what plants you want put in.'

The pub provided another release from the strains to which the sappers were subjected. It offered them some welcome normality and a sense of community, perhaps even a song around the piano. Many of the men spent their free evenings in the local, and while alcoholism was not common most sections had their heavy drinkers. George Ingram was among those who found that beer helped them to cope better. Although Ingram got used to the work, he suffered from recurring nightmares in which he was blown to pieces like Lieutenant Hoare. 'I used to say to civilians: "The slogan is 'Join the Army and See the World' – 'Join Bomb Disposal and See the Next One!'" . . . It did make you drink quite a bit.' Others in his section found their worries too much to bear. One sapper took the pin from a grenade and dropped it at his feet, only to be saved by a comrade who threw it away. Another man – 'He was a very nice fellow, he was courting' – who was fearful about being posted abroad did succeed in committing suicide. Even so, incidents such as these were so uncommon as to be memorable. Fear was natural and widespread; suicide was not, perhaps because there was the option to transfer out if the stress became intolerable.

Len Jeacocke was scared too. 'Some blokes just couldn't make themselves go down a hole one more time. They were put on a forty-eight-hour transfer and a replacement found immediately. As far as I was concerned, there was no shame. I was afraid myself. The great dread every time you approached another bomb at the bottom of an excavation was that it might have a delayed-action fuze, and might still be ticking.'

Yet when offered a chance to leave BD, Jeacocke chose to stay. As a senior NCO he was encouraged to apply for a commission and was told by his colonel that he would probably be accepted. On the basis of this, for a few weeks he went round with pips up, in effect an officer. Then he was told that he would have to go to an OCTU, and that afterwards there was no guarantee that he would be posted back to BD. Preferring to remain with the work and the men that he had got to know, Jeacocke withdrew his application and reverted to serjeant's stripes.

One of the dangers unique to bomb disposal was the camouflet. This was a chamber filled with the carbon monoxide created when a bomb exploded underground. Often there was no more sign of this than a slight hump in the ground, and the gas itself was odourless. It had a lethal affinity for haemoglobin that almost invariably proved fatal to those who fell through the thin crust above a concealed camouflet – so much so that rescue attempts were forbidden as they merely added to the casualties. Lifelines were supposed to be worn when working in shafts for this reason, but this instruction was often overlooked or proved impractical.

After languishing in Sheffield since 1939 with one of the original disposal trios, Harry Beckingham had joined 5BD at the start of the Blitz. His section was digging for a bomb in the centre of Ilford 'when out of the blue a German plane swooped down upon us, machine guns blazing as he roared past. I was in the excavation and the rest of our squad had taken shelter in nearby shop doorways. At this precise moment the floor of the excavation caved in.'

The next thing he knew, he was in a hospital bed. Beckingham had fallen into a camouflet, but fortunately a policeman had seen his head suddenly disappear and his pals had raced over and pulled him out, although not before he had breathed several lungfuls of carbon monoxide.

Some months later, William Wells's men were to have a similarly narrow escape from disaster. By then, a weekly rest day – usually Sunday – had been instituted for the sections. On a Saturday, the squad began to dig for a Category C UXB at Biggin Hill. The site gave every appearance of being that of an unexploded fifty-kilogram bomb. After knocking off for the evening, Wells returned on the following Monday and was 'horrified to see that most of the bottom had disappeared, revealing a roughly spherical cave about ten feet in diameter. It was a camouflet.

'The bomb of a hundred pounds' weight had penetrated the ground to a depth of about fourteen feet and being fitted with a short delay fuze had exploded and blown a cavity in the solid clay but failed to break the surface . . . The cavern looked harmless enough but we knew that it contained deadly carbon monoxide gas left by the explosion. Had any of the sappers been working in the excavation when the bottom collapsed, their death would have been inevitable and swift. Indeed they would have been overcome before a rope could be lowered and grasped.' They had been saved by the rest day.

By late September the sheer number of UXBs being reported, and the ability of the unproven bomb disposal organisation to master them quickly, was giving the government real cause for concern. Churchill had maintained his close personal interest in the struggle, and now sought to give BD's efforts fresh impetus. On 20 September, a week after querying the numbers of troops to be deployed on disposal duties, he wrote to the half-dozen cabinet ministers concerned with the problem in his newly created guise of Minister of Defence.

'The rapid disposal of unexploded bombs if of the highest importance,' stated his minute. 'Any failure to grapple with this problem may have serious results on the production of aircraft, and other vital materiel. The work of the Bomb Disposal Squads must be facilitated by the provision of every kind of up-to-date equipment . . . Priority 1 (a) should be allotted to the production of . . . equipment.'

Churchill was genuinely alert to the gravity of the threat posed by UXBs, having recently learned that even before the Blitz began there had been 724 Type 17s reported in the three weeks since the first was found. An especially heavy attack on Central London came on 18 September, with many of the stores on Oxford Street being smashed or closed by bombs.

Two further considerations also helped BD's cause. The first was that Churchill had much experience of the importance and the difficulties of military supply, having been Minister of Munitions during the First World War. For much of the thirties he had argued against the appeasers for its re-establishment. The second factor in BD's favour was his longstanding fascination with science, or at least with anything modern, especially if it seemed to offer a short cut to an objective.

One such enthusiasm pursued in these weeks was for augers. These were American-made drilling machines reputed to be able to bore in an hour a hole that would take several days to dig. Like many of the technical suggestions with which Churchill bombarded officials during the war, it was put into his head by his chief scientific adviser, the physicist Frederick Lindemann. Keeping track of the augers that subsequently had to be ordered – and then re-ordered after some were sunk in transit – generated a volume of official correspondence perhaps out of proportion to their usefulness, but the episode did demonstrate Churchill's ability to see clearly to the heart of issues. 'The essence of this business is to reach the bomb and deal with it with the least possible delay,' he wrote to Anthony Eden at the War Office. 'I consider that we owe it to these brave men to provide them with the very best technical equipment.'

Churchill's minute about the priority to be given to the production of this equipment led to a flurry of ministerial activity. A flustered Herbert Morrison penned a note to Gough the next day: 'DSR. Very urgent. I am not sure whether your responsibility extends to all these items, but will you please trace and coordinate and see that every effort is made.' The items covered by Churchill's order included several that, in the face of competition from rival ministries, Gough had had difficulty in getting made. Naturally he was delighted that their mass production was now to be expedited.

Among his priorities were the steaming-out sets, which had also been brought to Churchill's attention by Lindemann. On 26 September the latter had been given a tip that these were not being used on a wide enough scale, and within two days Churchill was demanding assurances from the relevant ministers that they were. Morrison had already attempted to mollify Downing Street earlier in the week – 'everyone is fully alive to the urgency of the situation' – but Eden was able to exploit Major-General Taylor's having a foot in Supply's camp to discover the true picture.

On 1 October Eden told Churchill that he considered the rate of production unsatisfactory. Churchill concurred. Within a few days Morrison had been moved to the Home Office, arguably a ministry that made better use of his experience as leader of the London County Council, and it was his successor at the MOS, Sir Andrew Duncan, chairman of the British Iron and Steel Federation, who wrote to the Prime Minister that 'We are alive to the urgency of the situation.' New firms were brought in to make the sterilising sets, which had also been simplified, and the expectation was that five per week could soon be rolled out for use in the field. The drilling time had been reduced to eight minutes, as steam pressure at 100psi drove a series of gear wheels that rotated spindles clockwise through anti-clockwise screw threads, giving greater control over cutting at just one thousandth of an inch per revolution.

At the end of September Merriman surveyed the state of BD's arsenal against the fuzes. There were some promising new developments in the pipeline, such as an improved key that would give greater leverage when extracting deformed fuzes. Professor Andrade's vacuum pump, which would allow liquids to be introduced into the fuze pocket, was also in production and might prove to be the answer to the (50), the motion-sensitive device. For the moment, standard procedure was not to touch these for two and a half days, after which Woolwich considered that it was 'definitely safe to handle'.

Trials were also ongoing of various methods for use against (17)s. The notion of injecting some kind of 'setting' liquid that would gum up its cogs was being followed up, while a hundred sets of the new magnetic clock-stopping coil had been ordered. For the time being, however, the fuze was still in the ascendancy and becoming an even greater nuisance; protected by the Zus, some bombs were lurking for up to six and a half days before exploding.

One pressing requirement was for a means of detecting UXBs buried underground. Gough and Taylor had spent the day before the London raids began testing the claims of Major Ken Merrylees, a member of the General's staff, that he might hold the solution to this.

Merrylees was a Royal Engineer through and through: so much so that a piece of shrapnel taken from his body after he was wounded in France in 1917 was found to bear the imprint of the RE badge (the splinter having struck his shoulder before entering). He was also a keen water diviner and dowser, and had told Taylor that he could find metal under the earth. Like many hard-headed military men, Taylor had a weakness for the fantastic, whereas Gough preferred more empirical evidence.

A series of six tests was arranged at the National Physical Laboratory at Teddington. In each, Merrylees was invited to point to a gas or water main that ran through the garden. His belief was that he could find any metal which conducted energy, using nylon rods to help him; latterly, he had had to employ these instead of whalebone, since women had ceased to wear corsets.

As Gough's waspish record of the day attests, the experiments were not a success. Merrylees could not find the pipes, and even when told where one was he asserted that it curved when in fact it was straight. Nor could he detect an actual bomb placed under the lawn. 'The only proper scientific deduction to be made,' concluded Gough, with almost audible satisfaction, 'is that the "dowsing" method is completely unreliable.'

Frank Martin had a choice to make. Five days earlier, four of his men had been killed while digging out a time fuze. Now 5BD Section was faced with another. It had landed inside Chevening, the family seat of the Earls Stanhope, near Sevenoaks in Kent. Lord Stanhope had been Leader of the House of Lords until May, when Chamberlain's government had fallen after the Norwegian debacle. Now his wife lay in bed, suffering with cancer, and the doctor's advice was that she would die if she had to be moved from the house. Yet Martin had heard the bomb's fuze ticking.

It had come to rest near the great Georgian curved staircase that rose from the hall. Rather than risk any other lives, Martin ordered his section outside and began to tackle the bomb alone. After several hours of solitary digging, at one o'clock on the morning of 17 September he succeeded in uncovering it. Although he knew that there might be a Zus behind the fuze, he took the risk that there was not and managed to extract it. He had gambled with his life to save that of the Countess, although unfortunately she was in any case to die of her illness the following day. Martin's actions were typical of the spirit with which BD was beginning to be imbued, as was the butterfly span of his time with it. A month later he was to be killed by another (17), returning to the fuze after the string with which he was trying to extract it snapped. The bomb had lain dormant for nearly four weeks. Frank Martin was twenty-nine, and the first of the original batch of sixteen section officers to die. His replacement was to be Ken Revis.

The changeable and often imperceptible line between safety and danger was one that all BD officers trod. Herbert Hunt was the first Divisional Officer of 1 Bomb Disposal Group, which controlled all the sections in London. A large part of his job was reconnoitring UXBs so that they could be assigned to sections: of the 650 that he inspected, forty-seven exploded spontaneously – sometimes just minutes after he had examined them.

Hunt had served with a London regiment in the trenches and, having been with the TA since, was called up in 1940. He volunteered for bomb disposal and was sent to Manby, where he was lectured to by Len Harrison. His most anxious moment there was a brief trip in a bomber, the first time that he had ever flown. Two of the dozen officers on his course were to be dead within weeks; a third was killed later in the war.

He had been briefly stationed at Pontefract, but was ordered down to Ilford two weeks after the Blitz began. Since he came from Upminster and knew London, he was then appointed reconnaissance officer north of the Thames, in which capacity he was eventually to log more than nine hundred UXBs, not counting incendiaries and anti-aircraft shells.

Much of his time was spent at bomb cemeteries, the places where UXBs were taken to be steamed out or to have their fuzes detonated if this could not be done on site. Wide, open spaces were needed for this. In London, Hampstead Heath and Hackney Marshes were used at first, while Regent's Park was resorted to in emergencies. Piles of bombs soon awaited disposal, while near by lay the gashed and twisted wreckage of those that had been dealt with.

Complaints about noise and damage to property from residents living near these sites were frequent and vocal. The London sections were eventually forced to move their operations out to

Richmond Park, Wormwood Scrubs and Chingford Plain. The most intransigent and parochially minded of all those approached for permission to use their land for such work were the owners of golf courses. It says something about the differences between British and German society that the necessary spaces were not simply requisitioned.

By comparison with other regiments stationed in London, such as the Guards battalions quartered at Chelsea Barracks and Birdcage Walk, life in BD was regarded as neither elegant nor comfortable. It had been decided that officers should be boarded out in case – as indeed happened on several occasions – their headquarters at the Duke of York's Barracks in Chelsea should be hit. Hunt was initially billeted with four others in a dilapidated house where the only washing facility was a footbath. A canteen was eventually installed at Chelsea, but until then meals had to be taken wherever they could be found. Given BD's long working hours, this made for an uncertain diet.

Sappers and officers alike got their hands – and usually feet, face and hair – dirty on disposal work. If they were not at the bottom of an earthen shaft, covered in mud and clay, they would be pad-dling in sludge in a sewer under a road, smeared with tar and worse. Battledress frequently had to be put on damp, day after day. Hunt had nothing but admiration for the fortitude and good cheer of the sections, which often had to toil by night, for instance manning pumps to keep a dig from slipping. This was done by the light of hurricane lanterns, that would need to be shielded from the rain by soldiers drenched to the skin, covered in mud and choking on the fumes from the petrol pumps. Despite all these incentives to complain, BD was, found Hunt, a largely happy family.

One of the compensations of the job was the variety of the work, which staved off the bane of most soldiering: routine and boredom. On one occasion, Hunt was talking to S. A. Smith, the CO of 5BD, in the company's offices in Chelsea when four young and doubtless attractive ATS officers came in to report a bomb on the roof of their headquarters. Three of Smith's subalterns immediately volunteered their services, but it was the middle-aged Hunt who soon found himself perched three storeys high, with one foot on the top rung of a long and none too stable ladder and his hands scrabbling for the coping above. When he succeeded in hauling himself over this he discovered that the 'UXB' was a large cast-iron mincing machine that had been flung up there by the detonation of a real bomb near by.

On one occasion Hunt had to search for a UXB in a graveyard, a task made all the more difficult by the lack of resistance from the layers of bones under the topsoil, which allowed the bomb to penetrate further than usual. Another time, early in the Blitz, he took a section to deal with a bomb that had landed next to a lunatic asylum.

Hunt descended to the bottom of the shaft and with wet clay tamped in place a charge to blow the fuze. As he was about to be pulled up there was a landslip and he was trapped by a great weight of earth and timber. Luckily none of the clods had hit the detonator or the ounce of dry gun cotton primer that he was preparing to fit into a slab of wet gun cotton: the first could have taken off his hand; the second would have mashed him to a pulp; while the whole charge was strong enough to blast through steel. Even so, not only could he not see daylight above him but now he could feel water around his feet.

Hunt had to wait for five hours while a rescue was prepared, the icy water creeping up around him inch by inch. When a rope was finally lowered the water had reached his chest and he was lapsing into unconsciousness, willing himself to stay awake. The first thing that he saw on reaching the surface was a crowd of the asylum's inmates, pressing forward curiously. Hunt was overcome by strange emotions and ran away as fast as he could, only stopping when he arrived at a pub.

In a trancelike state, he asked for two pints of beer: the first he downed without pausing; the second took him an hour, while he repeatedly counted his fingers. Looking back, he thought that he must have been trying to reassert control over a mind that had given itself up for dead.

Yet though such incidents could be grimly amusing, there was little relief from their constant worries for the families of BD personnel. This was brought home to the rather buttoned-up Hunt when in late September his wife was telephoned and told that he had been killed. In fact, his name had been confused with that of John Hunter, the officer caught in a blast at Mill Hill and a contemporary of Hunt's at Manby.

Hunt's wife, with whom he had a young daughter, had no means of confirming the news as he had been forbidden to give out his number at the Chelsea HQ as it was a line used only by the Police and ARP. The first that she knew that he was alive was when he chanced to ring her that evening from a callbox in Sloane Square.

'I understood how my wife must have felt at receiving such bad news,' he later recalled. 'I could do nothing but offer my sympathy over the telephone and although my home was but an hour's run from Chelsea, I could not be spared, as I was always on immediate call, day and night.' The best that he could do was to persuade a friendly operator to leave the callbox connected while she took cover during the bombing, so that he could talk to his distraught wife for a while.

A few days later Hunt went to examine a large UXB which had fallen at Westbourne Park railway and underground station. Only a few minutes after he had gone to investigate another report, it went off. Some air raid wardens had not seen him leave and informed his headquarters that there was no trace of him. This time, Hunt's wife was told by his CO that he had been killed. When Hunt learned what had happened, he could not bear to strain her nerves any further by telephoning himself. The only solution that he could think of was 'to contact a friend, asking him to go round and break the news gently that I was, in fact, alive'.

Managing his men's morale was an important part of every officer's responsibilities, but as William Wells appreciated, family problems could lead to lapses of concentration during work that was unforgiving of them. Years afterwards he recalled how one of his sappers had allowed news that his daughter was ill with pleurisy to prey on his mind. It was perhaps only natural for him to fret about her, and to be upset at the thought of her having to spend nights in a cold and damp air raid shelter. Then, while he was on the way to a site in Peckham where the squad had already been labouring for three days, Wells was handed a telegram by the section clerk. It was from the sapper's wife, and told him that his daughter had been asking for him. 'I felt like exploding,' he recalled, with perhaps unconscious irony, all too aware that the telegram could only serve to distract a man engaged in a hazardous business.

'The problem of sharing domestic responsibilities while husband and wife are separated through military service is not easily solved,' thought Wells. 'The wives of regular service officers and men learn to accept the fact that in a soldier's life duty is paramount, and family worries must in time of war be borne by women.' He put the piece of paper in his pocket as he neared the UXB, and resolved to hand it to the soldier when the job was done.

'As we stepped over the rope, there was a vivid flash and a deafening explosion. The houses seemed to rock, and above the bomb site there appeared a red haze, changing to a moving pattern of red and black smoke from which descended debris, human fragments and a dew of blood.'

In the early hours of the morning, Wells remembered the telegram. The sapper had been among those killed, so he consigned it to the fire.

11

FOR GALLANTRY

The mood in the pub was sombre. That morning, 16 and 17 BD Sections had been to a funeral for half a dozen sappers from 60 Section. All three squads were based at Bunhill Row, and they knew the dead men well. They had been killed while digging for a UXB near the railway line in Dalston, by what the report termed a 'premature bomb explosion'. There were no witnesses, and no survivors.

Somebody turned on the radio to listen to the news, and George Wyllie was startled to hear his name. Then Lieutenant Davies was mentioned. It seemed that, with a civilian, Thomas Alderson, they were to be the first recipients of a new honour for bravery, awarded for their actions at St Paul's two weeks earlier. The announcer said that, as mentioned in the King's recent broadcast, the medal would bear his name – the George Cross.

The force behind its creation, and its subsequent association in the public mind with bomb disposal, was once again Churchill. The original impulse to reward courage by civilians, however, was that of one of his closest advisers, Major-General Sir Hastings 'Pug' Ismay. As Chief of Staff to the Minister of Defence – Churchill – Ismay acted as a bridge between the military and

civil leadership during the war, and in mid-August 1940 suggested that the civil gallantry awards might be overhauled to take
account of the German air offensive against Britain.

By the start of September, Churchill had seized on the idea of
a 'Home Defence Medal'. His original vision was that it would
encourage factory hands to continue with vital war work during
air raids. As the attacks on London began, however, he told the
War Cabinet that 'the civil population was now in the front line'
and he decided to broaden the new honour to include the police,
ARP and emergency services.

'At the moment,' he told the Home Secretary, 'I am inclined to
a "George Cross", which may gradually come to hold its own
with the Victoria Cross.' He had discussed the medal with the
King, who liked the idea, but Churchill had not mentioned his
proposal for its name: 'I thought we might announce the new
decorations in the [King's] broadcast in the next few days.'
Meanwhile, Eden informed the Cabinet that the War Office was
having difficulty in adequately rewarding the bravery of BD
squads. Since their deeds were not 'in the face of the enemy', they
fell outside the criteria for medals for valour, and he proposed
awarding them the Empire Gallantry Medal (EGM) instead. The
first four – including Edward Talbot and Jack Button – were to be
gazetted the next day, but it was agreed by the Cabinet that the
EGM was not well-known enough for the recognition they
sought.

So it was, at that very instant, that the media storm inspired by
the saving of St Paul's burst upon the government. There was a
clamour for the section to be given the VC, letters to editors
querying the rules surrounding it, and story after story about the
hero of the hour: 'A man the world is talking about – Robert
Davies, time-bomb killer.'

Clearly the high-profile nature of the squad's actions made
them the ideal candidates for the new decoration. Giving the St
Paul's section the first George Crosses would firmly establish the
existence and purpose of the award in the public's mind, and

encapsulate the idea that everyone was pulling together on the Home Front.

On 30 September, therefore, the news broke that Davies, as the officer in charge of the working party, and Wyllie – the sapper who had uncovered the bomb – would be the first two soldiers to receive the GC. Alderson, an air raid warden who several times had risked his life to rescue people trapped under wreckage, would be the first civilian GC. Meanwhile, the determination and devotion shown by Davies's sergeant, James Wilson, and by Lance Corporal Bert Leigh, who had helped Davies finally to dispose of the bomb, was to be recognised with the British Empire Medal.

The announcement raised the hullabaloo surrounding Davies to a still more feverish pitch. The *Daily Mirror* conveyed news of it to his wife Isabella, whose reaction confirmed the image of him as a modest everyman.

'Well, I reckon he deserves it,' she told the paper, 'though it won't make any difference to him. We can't think any more of him than we always have. He has always been the best man in the world to us.' After learning of the honour, Davies wrote to his wife in similarly self-effacing terms: 'It seems as if they are trying to make me a national hero, but don't take any notice of it. I am still the same old dad.'

A few days later in Cardiff, Talbot was to receive a letter authorising him to exchange his newly awarded EGM for the GC. Stuart Archer, his fellow section commander, was not feeling quite so on top of the world.

Two nights after his recovery of the first Zus at the refinery, he and his men had finally finished defuzing a stick of ten fifty-kilogram UXBs in the shunting yards at Swansea. The operation had taken all day, even though these bombs had been dropped at low level and had not had time to arm themselves. Archer began to drive the lorry through the dark, back to their barracks with the section flaked out in the back, sprawled among the bombs.

The blackout was in force, even for vehicles, and so when he came over a hill Archer did not see a truck that had stopped on the road. It had been hit by a car, and two people had been killed. Archer's truck now rammed into the back of it, breaking his leg. In great pain, he was taken by car to hospital in Cardiff, but it was not until the next day that another officer knocked on the door of his house to let his wife Kit know what had happened.

She had passed an anxious night when her husband had not returned, and the officer began tactlessly by saying that he had some bad news. It is indicative of the strain she was feeling that when he then revealed that Archer had broken his leg, her reaction was a joyful shout of 'Thank God for that!'

Archer's injury developed complications, and it was not until four and a half months later, in mid-January 1941, that he resumed full duties. In the interim, a number of changes had begun to be made to BD's working habits, now that they had more experience of their foe. The most immediately useful of these concerned the recognition and reporting of UXBs.

By early October 1940 the categorising system was in place and paying dividends. The ARP organisation was also better dragooned by the commissioners who controlled civil defence in each region. In particular, they were reporting fewer false alarms, as selected wardens and police officers were trained to identify the presence of UXBs from the appearance of holes and craters.

An exploded bomb weighing a thousand kilograms could form a crater up to fifty-two and a half feet wide and sixteen feet deep, with a fuze that had gone off underground leaving a larger scar than one that had detonated on impact. The size of a potential UXB was judged from that of the entry hole. At eight inches it was probably a fifty-kilogram bomb, and at eighteen inches up to five hundred kilograms. How far it had penetrated would depend on the height from which it had fallen, its weight and the nature of the soil on which it had landed: rock was the hardest, wet clay the least resistant, allowing a bomb to sink down as far as sixty-five feet.

Estimating the size and type of the bomb was crucial to determining safe evacuation distances and times. For instance, (17)s were usually found in 250-kilogram bombs, so if a buried UXB was thought to weigh fifty kilograms, and thus probably had only a dud impact fuze, it would be safe to allow the local population to return to their homes after one day and not four. Traffic, however, would still need to be kept at least a hundred yards away to avoid setting up vibration in the fuze.

It was also vital to confirm whether a bomb had exploded above ground or was in fact a UXB below it. It was not always easy to tell. A half-ton missile that had slammed into a house at 500mph might demolish much of it before concealing itself beneath the debris. The best test was to see if the glass in the windows had been broken: this was evidence of their having been a blast. Other indicators included circular rather than radial cracks in the earth, the smell of fumes and fragments of the bomb itself.

The impression that BD was starting to get a grip on the situation was met with approval in Downing Street. On 9 October Churchill wrote to Ismay that, 'We have not heard much lately about the delay-action bomb, which threatened to give so much trouble at the beginning of September. I have a sort of feeling that things are easier in this respect . . . Is this easement which we feel due to the enemy not throwing them, or to our improved methods of handling?'

The probable answer is a little of both. Although the focus tended to be on the (17)s as the most disruptive fuzes, they did not represent the dominant proportion of bombs dropped. At the end of September a proper log had begun to be kept of each UXB reported, and the statistics collated from these buff forms would show that fewer than 10 per cent of all bombs exploded some time after they had landed, and that the bulk of these were (17)s. More than two-thirds of the German bombardment consisted of fifty-kilogram SCs or general-purpose bombs, the fuze pockets of which were too short to hold a time fuze. It was these that made up the majority of UXBs.

Yet whatever their nature, unexploded bombs of all kinds were bringing Britain to a halt on a scale far in excess of that with which BD could yet deal. In the company into which Archer and Talbot's squads had been newly absorbed – No. 8 – UXBs were defuzed at a rate of 0.73 per day per section, which still left a daily average of 4.36 bombs per section outstanding. Many of these were not urgent cases, but if multiplied by the 440 sections now at work it was enough to give Taylor palpitations.

Another measure of his task was supplied by the MOHS on 11 October. The ministry had been sifting through the buff forms for the Cabinet and reported that some 3830 UXBs had been reported in the three weeks since the attacks had begun in earnest – or about 150 new sightings per day – with much the greatest number being in London. Many had been already tackled, only for fresh ones to take their place, so that there were presently 3225 unexploded bombs awaiting disposal. Of these, 853 were in the capital, and a hundred of those were rated Category A, a major and immediate threat to the war effort. Where there had been just twelve sections in London, there would soon have to be eighty-eight.

Statistics compiled after the war showed that approximately 8.5 per cent of all bombs dropped became UXBs. They also revealed that Churchill's satisfaction was premature. The outlook may have appeared rosier, but by far the greatest weight of the Blitz, and of UXBs, was still to come.

One improvement that gave hope of reducing the ever-mounting total was the establishment of Divisional Officers, such as Herbert Hunt in London, charged with maintaining better liaison between BD and Civil Defence (CD). Hunt did much of his work for CD Group 7, which looked after twelve boroughs and fifty-three thousand acres in east and north-east London. Its network of ARP wardens made the Divisional Officer a natural point of reference for local sections, while his remit also extended to dealing with district engineers and surveyors, as well as the sourcing of pumps, cranes, compressors and other useful non-military machinery.

Public awareness of BD and its work had also increased. The red-painted mudguards of their lorries – intended to speed their passage through police cordons – were becoming an ever more familiar sight, and public appreciation of the risks taken by the sections rose accordingly. Cups of tea were the least welcome that they could expect, and when Christmas Day came the whole of Jim Lacey's squad would find themselves invited out for lunch by their neighbours.

An early perk of the job was selling on pieces of UXB as souvenirs, but a stop was soon put to this. Accepting money collected by grateful residents was also banned, although it was difficult to police. William Wells knew of one officer who had been approached by the owner of some flats with the offer of twenty pounds (about £850 now) to remove immediately a bomb that was not a priority: its presence had forced out the man's tenants, and as a consequence he was receiving no rent. In this case the officer refused, but he told the landlord that he would be happy to distribute a gift if it was still forthcoming once the bomb had been made safe. Wells thought it 'right and proper that the sappers should accept these rewards', which were usually sums rather smaller than the twenty pounds proffered in this instance, provided there had been no solicitation.

Any BD officer seen in a pub was liable to find more pints than he really wanted being bought for him. Other comforts that came their way included free membership of the Royal Automobile Club in the West End, which with Kempinski's restaurant in Piccadilly became the unofficial base for any section officer lucky enough to have an evening off. Another bonus was that by November even the most junior subaltern in BD ranked as a lieutenant; the accelerated promotion was agreed by the government so as to head off talk of 'danger money' being paid to the squads.

Soldiers were sometimes tempted to make up for the absence of this financial reward in other ways. Looting from bombed-out shops and houses was not unknown, as was taking advantage of requests from owners to enter cordoned-off premises to fetch money or jewellery. Two sappers in Archer's company were imprisoned for abusing this trust. They were overheard discussing their haul in a dockside pub by undercover field police keeping an ear open for loose talk about Atlantic convoys.

Examples of dissatisfaction with a squad's work were so rare as to be collector's items. John Emlyn Jones, who was later to command 5BD Company, recorded one such, which occurred when a crew was digging for a Category A bomb in the West End. 'Being no different to anyone else in khaki', they had downed tools when a YMCA refreshment car appeared and settled back to enjoy a mug of tea and a smoke. Five minutes after they had left the site, the UXB exploded. The only person not to see this as a lucky escape was the irate owner of a building wrecked by the explosion, who complained in unambiguous terms of 'the lack of responsibility and devotion to duty on the part of the working party'.

Despite the hazards presented by bombs morale in Bomb Disposal remained high even as the Blitz worsened. 'I think,' explained Wells, reflecting on the few requests made for a transfer, 'the men reasoned that bomb disposal was essentially a sapper's job and working conditions were no worse than in other sapper duties: there was a minimum of parades and a pride in doing a very useful job of work.

'The men seemed unaffected by the danger and by the heavy casualties which occurred in the early days . . . I found that only when men had witnessed a blow-up were they affected. It frequently happened that the driver of a truck was the only survivor of a working party, due to the fact that when a blow-up happened he was engaged on daily maintenance tasks on the truck, parked a safe distance from the bomb. Then the driver had the ghastly experience of seeing his companions blown to pieces. I knew of two such cases. The Medical Officers responsible for the section allowed the men to return to their units, but in both cases the men's nerves were so affected they could not resume driving duties.'

Those profoundly affected by witnessing a disaster might be offered a transfer, but often they preferred to stay with the section, where they would receive sympathetic treatment from others. 'I came to an early conclusion,' Wells concluded, 'that the companionship in danger contributed in a large measure to the success of bomb disposal in the early months of the Blitz, when new fuzes and bombs were being used and the factor of the unknown was always present. In these months when casualties were highest, so was the morale of the men.'

As befitted a young gentleman of independent means with apartments in Jermyn Street, Freddie Leighton-Morris was used to getting what he wanted. When the war began he had tried to join the Army, but as doctors had only given him four years to live on account of 'a groggy heart and wonky lungs', he was turned down. Even Bomb Disposal refused him. So when he returned home one day to find his way barred by a constable, who told him that an unexploded bomb had landed in the rooms next to his, he saw his chance to do his bit.

Initially he volunteered to remove the bomb, but was told by an inspector that he would be arrested if he entered the building. Leighton-Morris was not to be so easily thwarted. While the police were busy at the back he strolled in the front, let himself into his flat and climbed up the fire escape to the bedroom where the fifty-kilogram bomb had been found. Cradling it in his arms, he then began to stagger as fast as he could wheeze towards St James's Park. At one point he dropped the UXB on his foot. Soon afterwards he was arrested by a somewhat nonplussed police sergeant.

At the Bow Street courts in early November the magistrate praised his 'extraordinary courage', but said that only 'those in authority' could 'decide in what part of London a delayed-action bomb should go off'. Leighton-Morris was fined the substantial sum of one hundred pounds for contravening a police order. The story, however, had a happy ending: when the matter was raised in Parliament Churchill ensured that the fine was reduced to just five pounds and a few weeks later Leighton-Morris was passed fit for service with an anti-aircraft unit.

Most Londoners were content to leave bomb disposal to experts such as Bob Davies. Some of the shine had been taken off his exploit at St Paul's with the news that the same week four of his men – an NCO and three sappers – had been accused of stealing from a burned-out shop. No doubt this could be ascribed to the strain they had been under, as could the conduct of two other of his sappers who had been arrested for assaulting a pair of special constables outside a pub. His own lustre, however, gleamed ever more brightly.

In late September he removed a bomb from the British Museum, and after dealing soon after with another at the Royal London Hospital was presented by the staff with a stethoscope. He had frequently borrowed one from them, related *The Times*, to 'listen in to delayed-action bombs' before defuzing them. The surgeon who made the presentation said that after the UXB had fallen they had first 'called in "the local doctor", and then "the specialist" – Captain Davies.'

Indeed, so synonymous had he become with BD that it was his name that was given by one Victor Langston, a fitter from Lisson Grove, to convince police and civilians that he was a member of a bomb squad, allowing him – a court heard – to scrounge a car and petrol. More seriously, Langston had told the owners of a café in the West End that the UXB within it was safe when it was not.

The dedication of Davies seemed to know no bounds. Even when given a week's leave following his commendation for the GC, he was to be found perched for the photographers atop a bomb being excavated at – of all ironies – the German Hospital in Hackney. The presence of the press and the squad's rather staged poses suggest that the bomb had already been made safe, but the captions left readers in no doubt that this was just how it had been at St Paul's: 'After the fuze was removed, it was transferred to a lorry – like the one at St Paul's – and taken to Hackney Marshes to be detonated. Then Lieutenant Davies and his men calmly got on with their next job.' Bob Davies had become BD's poster boy.

Later that week, on 9 October, he was made Divisional Officer for his area. In the last fortnight of September, 5BD had disposed of an estimated six hundred UXBs. In other words, its sections had each dealt with an average of two bombs every day, all of them a threat to their own lives and those of people nearby. And there were still another 632 known UXBs in the capital. The Company War Diary for 30 September states that it cannot give a more precise figure owing to the 'dearth of Section records' before the middle of the month; the time of the entry is given – wearily – as 1760 hours.

The cost of pressing on would continue to be paid in blood. The afternoon of Thursday 10 October was quiet, and Hunt made the most of the opportunity to treat himself to tea at a restaurant in the King's Road, Chelsea. He was joined by Second Lieutenant Lionel Carter, a twenty-year-old newcomer to 2BD who was helping one of the more senior officers, Oswald Robson, and who was keen to make a good impression. For a few minutes chintz and china made life seem almost normal again.

Carter mentioned that a little later he was going to the bomb cemetery at Regent's Park. A 250-kilogram had been brought into the Duke of York's Barracks and ought to be blown up. The fuze was unidentifiable, having been badly distorted when it had passed through a cast-iron drainpipe. It had proved impossible to remove, though self-evidently it could no longer be working. He wondered if Hunt wanted to come along, perhaps to show him the ropes, but Hunt was keen to get home. So was the driver in 61 BD Section – Cecil Brinton's squad – as he had a date for the evening. Sportingly, Carter agreed to drive the truck himself, and to buy the men a drink afterwards. It would be a good chance to get to know them a little better.

Five of them heaved the dead bomb into the back of the vehi-cle and piled in alongside it. They had reached Madame Tussaud's in Marylebone when it exploded, obliterating all six men and the lorry. Somewhere along the way the fuze had re-started. It was the second decimation of Brinton's section in a matter of weeks, and by the time that he left 61 BD after fifteen months' service, only five of its original strength of thirty were still alive. Other men had taken the places of the dead, but he could never forget the experience of picking up the remnants of his friends so that their families would have something to bury.

Shortly after midday on 17 October Davies's three sections were drawn up on the parade ground at City Road when an ear-splitting roar burst upon them. From the direction of Shoreditch, a couple of hundred yards away, a tower of smoke began to stain the sky. As one, the officers and men ran towards it. What they saw when they arrived almost defied belief.

Six whole streets had been virtually levelled to the ground. Where a few minutes before had stood long sooty terraces of two-storied houses – homes that were not very grand but made less drab by pot plants and ornaments and photographs of sons in the Forces – now there was only rubble. The most fearsome weapon of the Blitz, a parachute mine, had gone off and some-where beneath the vast heaps of brick and slate and rafter were the two men who had been trying to defuze it.

Originally designed to sink ships, the distinctive cylindrical mines had begun to be dropped into coastal waters by parachute since November 1939. A Heinkel HE-11 bomber could carry a pair of the larger Type C magnetic mines. These were about nine feet long and two feet wide, and weighed about two tons, of which almost half was Hexanite explosive, similar to the British RDX and significantly more powerful than mere TNT.

The clocks of the mine's fuzes began to run on impact with the sea, to prevent them being recovered if they floated, but they were brought to a stop by water pressure once they were below two fathoms. There they waited until a vessel passed over them. In the early months of the war naval teams at Portsmouth began to study those that washed ashore, although not without cost to life; in August, six sailors were killed when a Type C exploded while it was being dismantled at their HMS *Vernon* base.

The mines became a still greater menace from mid-September, however, when the Luftwaffe started to use them as bombs against London. Drifting down on parachutes, their flight could not be targeted and their blast – often occurring at roof-top level – was indiscriminate. To the watching bomber crews, their effect appeared spectacular because the high proportion of aluminium in Hexanite produced a vast bright flame. This caused them to over-estimate the damage that they did – paradoxically leading to an increase in their use – but even so their destructive force was for-midable. A stretcher-bearer caught when one exploded in Pimlico on the site of a rescue operation remembers that it was 'just as though a huge orange flare had gone up under your feet. A hell of a bang. Then it was like a sandpapered ramrod down your throat, and your lungs puffing out like a pouter pigeon. Then dead, dead silence. Then, as though some time afterwards, a slow shower of bricks from everywhere.'

Delivered as they were, however, the mines' parachutes were often snared by stanchions or snagged by guttering, and so the fuzes – which jammed easily after a jerk or a blow – failed to det-onate. The unexploded parachute mine (UXPM) became a grave new problem, requiring as it did every building within a blast radius of 1200 feet – a zone amounting to over a hundred acres – to be evacuated. On the first night that they were dropped, that of 16 September, seventeen out of twenty-five mines landed intact, most with clocks that had started and then stopped, and by early October almost two hundred UXPMs had descended on London.

The first counter to these was an improvised device that mim-icked the effects of water pressure. It consisted of the rubber bulb of an old-fashioned car horn connected to two lengths of brass tube. The end of one of these was attached to a bicycle pump, while the other was clamped over the fuze. When air had been forced into the bulb a tap was opened, pouring the air into the mechanism and triggering a plunger that blocked the clock. By 4 October the 'safety horn' had been used successfully to clear London of mines. Nevertheless, there was always a risk that any movement of the mine would jolt the fuze into life and, as was soon learned, the clock had a maximum running time of just seventeen seconds. The work was carried out by a band of a dozen officers, each teamed with a rating. Ten of this group of volunteers would win the George Cross or the newly instituted George Medal.

As HMS *Vernon* was still concerned with mines found at sea, the Rendering Mines Safe (RMS) section of the Department of Torpedoes was set up by the Admiralty in late September. Its remit was to deal with those mines that fell on land, and it was supervised by Captain Llewellyn Llewellyn.

A former Chief Inspector of Naval Ordnance who had entered the Fleet in 1893, Llewellyn had been retired for more than a decade when in 1940 he was brought back to the Navy to be its first Director of Unexploded Bomb Disposal (DUBD). As Herbert Gough discovered at meetings of the Unexploded Bomb Committee, and in their regular correspondence, Llewellyn was an astute and combative figure who could rarely be prevailed upon to compromise what he saw as the Navy's best interests.

Then, on 15 October, when the moon was at its fullest, the Germans made their heaviest night attack so far on London. Perhaps as many as four hundred bombers savaged the city, killing more than five hundred people and badly injuring another thousand. In addition, nine railway stations were severely damaged, along with the Beckton Gasworks, Battersea Power Station and the Fleet Sewer. More than eleven thousand people were left without homes.

The night also signalled the start of a second onslaught of parachute mines. One fell the following evening through the roof of a house at Clifton Street, close to Finsbury Square in Shoreditch.

The next morning, before breakfast, Sub-Lieutenant Jack Easton of the Royal Naval Volunteer Reserve was sent to have a look at it.

A third-generation solicitor in peacetime, Easton was already as much a veteran of UXPMs as it was possible to be. He had recently defuzed one that had been hanging from a chandelier in the Russell Hotel, Bloomsbury. The proprietor had been so pleased that he had given Easton a cheque for £140 – about £6000 now – and promised him and his family free Sunday lunches there for life. Rather to Easton's chagrin, his CO had torn up the cheque, telling him, 'We do this for honour, not money', and had forbade him accepting a lunch.

With Easton at Shoreditch was Ordinary Seaman Bennett Southwell, who at twenty-seven was by ten years the younger of the pair. His task was to act as an extra pair of hands while Easton dealt with the fuze, and so together the two men set off down the abandoned street. That kind of solitary walk, wrote Easton later, always reminded him of the end of a Charlie Chaplin film.

They entered the house cautiously. To their right was the door to the parlour, which was below the hole in the roof, but when Easton eased it open it stuck fast after a few inches. Rather than force it, he made his way down the street and across some garden walls, then clambered through the parlour's back window.

Swaying gently in the centre of the room was a Type C. Its parachute had wrapped itself around a chimney, and as the mine had plunged through the house the lines had also caught on an iron bedstead. Its bulbous nose was now resting on several planks that it had thrust down and which had blocked the door from the inside. Easton had no choice but to tackle the mine where it dangled, and there was no easy escape route should anything go wrong.

Southwell began to pass him tools through the half-open door. It was dark in the parlour, so it was by the light of a candle in a cup that Easton started to fit the safety horn, but the fuze had been distorted by its passage downwards and the device kept

slipping off. Staying as calm as he could, Easton handed the useless kit back to Southwell and tried instead to unscrew the keep
ring of the fuze. Yet even when he applied all his strength it
would not yield. He had been trying this for about a minute
when the bomb suddenly began to slide to the floor. The next
two things he heard were the noise of the falling chimney pot,
which had snapped under the mine's weight, and the whirr of a
clock. 'Unless I got clear,' he realised, 'I had exactly twelve seconds to live.'

Yanking the door open, he was out of the house in two
bounds. From the corner of his eye he saw Southwell running
down the road, but Easton had no time to follow him. He threw
himself behind a brick air raid shelter opposite the house. 'I heard
no explosion. It has since been explained to me that if you are
near enough to an explosion of such force unconsciousness is
upon you before any sound reaches you, which is a merciful
thing. I was blinded by the flash that comes split seconds before
the explosion, but that was all I experienced.'

When he came to he could barely breathe, deep as he was under
rubble. 'My head was between my legs, and I guessed my back was
broken, but I could not move an inch. I was held, imbedded.'

An hour later he was dug out by Davies's men, and only just in
time. Fifteen minutes afterwards the building under which he
had been lying collapsed further. Of Bennett Southwell there was
no trace amid the devastation that had wrecked scores of houses.
His remains were not found until six weeks later. He had been
decapitated by the blast.

Jack Easton was still in hospital in late January when the news
came through that both he and Southwell had been awarded the
GC. As the rest of the ward broke into cheers, from under his bed
the nurses produced a bottle of champagne that they had been
keeping ready for the announcement.

Two days after Easton's rescue, Davies and Chadwick were summoned to deal with a Category A UXB in the basement of a

building in Lombard Street, in the heart of the City. The 250-kilogram bomb lay under rubble and neither of its two fuzes was identifiable. The standard combination would be a (17) and a (50), making it risky to move even after several days, but in the circumstances the sappers had little choice but to chance it. When the bomb was brought up to street level one of the fuzes was found to be a ticking (17), and it was rushed to Hackney Marshes, where it exploded three hours later. The Lombard Street building housed the offices of Charrington, the brewers, and its managing director was so grateful that he presented Davies with a most generous thank you, a cheque for five hundred pounds.

This bomb was to be one of the last dealt with in London by Davies. His reputation was now such that he was to leave for Cairo to lend his expertise to Middle East Command, which was encountering much unexploded Italian ordnance following Mussolini's recent incursion into Egypt. There was still plenty of work, however, for the Bunhill Row sections. In October 5 BD Company defuzed 698 UXBs. They had been able to discount reports of another 754, but this still left a backlog north of the Thames of some two hundred unexploded bombs. Half of those made safe were British anti-aircraft shells, but the great majority of the backlog were of the small fifty-kilogram type, and only one was as big as that at St Paul's.

Then, in mid-November, all three City sections found themselves confronted by the largest bomb yet dropped on London. At 1800 kilograms, it weighed as much as five Austin 7s, and it had landed right on top of the GPO's main sorting office at Mount Pleasant. The task of digging it out began at two o'clock on the morning of 16 November, a few hours after it had fallen, and continued for the next week and a half. Parties from 16, 17 and 60 Sections worked in shifts, day and night. One of those involved was George Wyllie, who remembered it as the worst job that he had during the war, because the bomb was just below the Underground track.

As ever, all the while that the sappers were labouring there was the risk that the UXB might explode, especially during the first three days. After that it was less likely that it would prove to have a time fuze, and indeed when it was finally recovered it was shown to have a dud short-delay anti-shipping fuze, like those found by Archer and Talbot in Wales. Not that that made the bomb any less monstrous. It was the largest type in the German armourers' arsenal — almost nine feet long, even without its fins — and had been nicknamed by them 'Satan'.

Elsewhere, some of the destruction wrought by its smaller brethren was starting to be cleared. Five thousand men of the Pioneer Corps had arrived in early October to help shift a third of a million tons of debris from central London. Good brick would be re-used to build shelters; steel would be salvaged for Spitfires. The rest would be dumped on the Essex marshes. Twelve mechanical cranes had been ordered from America, as had odd-sounding new machines called bulldozers. Slowly the skeins of civic life were beginning to be woven back together. Shattered gas mains could be repaired, electricity cables replaced. Britain's capital was showing itself to be more resilient than some had expected. Even the animals at the zoo had learned to cope with the bombs: when the raiders came over, the monkeys took shelter in the inner fastness of their hill, while the zebras shrugged off a blast that peppered their house. London, it seemed, could take it.

The first time that Bert Woolhouse heard of Bomb Disposal he was on a train from London with his friend Jackie Lewis. They were going up to Hull for a week's training for something or other. Jackie showed him an article in the paper about Davies and the St Paul's UXB. Cor, thought Bert, well fancy doing that! When they got back to London it was Jackie's birthday, and they went out on the town on the proceeds of Bert's winnings at the card school. They ran into a fellow in the Navy who had been at school with Bert, and it was all a bit of a blur after that. They had got drunk on rum, and Bert could remember falling down the barrack-room stairs. The next morning, nursing terrible hang-overs, they reported to Ilford for BD duty.

When they had arrived in Hull they had discovered that their 'training' was in fact a posting to 33/34 BD Section. Bert had been rather upset at finding that out, but what had made it all right was Mr Ash. Lieutenant William Ash was the most won-derful man that Bert had ever met. 'He was only a short fellow with a bald head, about forty-five' − he was in fact thirty-nine − 'and he had eyes like two ice-blue chips. You felt you could trust the man.' When Ash had learned that Woolhouse was not a vol-unteer he had told him that he did not have to stay, but Bert had decided to remain. 'We've got to,' he had said, 'we've got to join the Army to do something.'

Bert had just turned twenty-five when he joined up. He had thought the war was not going to last long. He had got married the week before, to Jane, whom he had met when he went with his brother-in-law to do a painting job at the house where she was working as a domestic. Within three months they were engaged. Bert was a decorator by trade, though he would have liked to have been a sportsman. His father had been a foreman, but he drank and had made their home life in Chalk Farm miserable for Bert's mother. He had adored her, but she had died when Bert was eighteen.

For the first ten days they did not see a bomb. There were five men in the section: Slim Wild, the driver; Bert and Jackie, who was a bricklayer from Chester; Swannie, the carpenter; and Titchie Websdale. Titchie was only twenty-one, but 'he was fearless, like a mole in a hole, he used to work in the circus putting up the big top'. He liked a drink, too, and after a couple of pints they had to get him out of the pub or he would start a fight with the biggest chap there.

The bespectacled Bert was not fearless, but he tried not to show it. 'When you're on bomb disposal, the last thing that you would ever let anybody know was that you were frightened. And I can assure you that I used to be scared out of my life.

'When you're in a hole, the other people are not with you, you're entirely on your own for twenty minutes, digging. You don't know what you're going to find. All my past life, all my misdemeanours and everything. I used to think about all of my most pleasant things to stop myself being scared out of my life.

'If anybody turns round and says they weren't scared, at least you could turn round very politely and say you were very apprehensive. Well, being very apprehensive is just being short of being scared. Of course we were frightened. I never sat on bombs and expected to come up smelling of roses. I expected something to happen.'

All day they scoured Ilford and Romford for UXBs, then at night they drank. Ash knew full well what they were up to and would ask to which pub they were going. 'And five pints of beer would come across, we'd all be playing darts, and we'd be told it was from that gentleman over there. And there was Sapper Ash. He'd put his stick to his cap, he was keeping an eye on us. And that's what Mr Ash was. He was keeping an eye on us.'

After a fortnight or so on bombs, Bert found a way out: a course to become a staff-sergeant. On his last morning, 7 October, he went to the section office to pick up his travel warrant. Ash told him that earlier in the week there had been a direct hit on an Anderson shelter at Connor Road in Dagenham, which had killed a honeymooning couple. He wondered if Woolhouse could have a quick look at what he thought was probably an exploded ack-ack shell. Bert was just to satisfy the civilian authorities and then get off to the station as soon as he could. And Ash told him that it had been a pleasure to serve with him.

Titchie went into the shelter first. It was not a pleasant sight inside, what with there being bloodstained rags still on the ground. The four of them each wrote their number on a piece of paper and shook them up in their hands before choosing one, to determine the order of their turn to dig while the others took cover. When they had got about four feet down they found a fin of a bomb unlike any they had seen before, bright green with a red rim, so they sent for Ash. They uncovered the first fuze, which was a (17). The second one was a (50), and as soon as they saw it they scrambled out of the hole as quickly as they could.

Almost five days had passed, though, since the missile had fallen to earth, so as far as was known it was safe to proceed. Ash sent for shear legs and tackle, and a tripod was erected above the thousand-pound bomb. Some other men arrived and, after Ash had listened for ticking and heard nothing, he tied the rope with a double clove hitch and gave the shout to heave.

'Titchie and Jackie jumped on one leg,' remembered Woolhouse, 'Ash and his driver on the other so it didn't go in. They were laughing because I got the job of pulling on the rope.

'Well, the last words of Sapper Ash were: "Don't stop pulling, keep pulling and pulling, it's very important!" And of course we'd never done anything like this before, that's why . . . We'd got the bomb out of the ground, we were holding it, we was told to hold, when we were holding it there was just this blinding flash. *A terrifying experience.* You see, it could never happen to you. It was other people that got blown up, you never got blown up. I remember my glasses flying away, and I was saying "Oh no, oh no".'

Bert had been hauling on the rope with half a dozen others. What probably saved his life was that when the bomb exploded the sudden absence of the weight that they had been lifting sent them all sprawling and the blast passed over his head. Some were less lucky: 'All they found of Jackie was his leg in a wellington boot. That's what they buried . . . One bloke, Pinkie I think it was, got his arm blown off. Swannie was badly injured, I was the next one up. I remember getting up and something hit me on the head. It was a chimney pot. There was rubbish flying all over the shop. You don't know what's happened to you, your mind is scrambled.'

Slim Wild had been one of those helping to hoist the bomb, and had been blown about thirty yards. Although badly shaken, he began to look for those who were missing. It transpired that five men had been killed: Lieutenant Ash and Second Lieutenant

Leslie Foster, the officers' driver Leslie Hitchcock, Jackie Lewis and Titchie Websdale. Seven sappers needed hospital treatment for their wounds, including Woolhouse, who woke up in a ward and assumed that the women dressed in white and ministering to him were angels.

The cause of the blast remained unclear, though several theories were put forward. It was thought most likely that the tug of the rope had re-started the clock of the (17). Herbert Hunt, who was to succeed Ash as CO of what was left of the section, records in his notes that a clock-stopper was used on the fuze, and speculates that it was the shock of it being applied which had set the wheels turning again. In his evidence to the inquiry, however, Slim Wild made no mention of the machine being used, and was adamant that Ash had listened for any ticking. Statistics would ultimately demonstrate that though such an event was much feared, in fact once jammed only about 1 per cent of (17)s ever re-started.

Ash had dealt with fourteen UXBs in a little over a fortnight, so was not inexperienced. Given what was later known about the increased longevity of the (50) fuze when buried in cold ground, it may have been this that was in fact responsible. Certainly, the incident led to a recommendation that a preliminary jerk be given to any bomb before it was moved. Yet the truth was that no one could be sure what had happened, nor how to ensure that it did not happen again. The only solution was more research, whatever price had to be paid.

When Jane Woolhouse opened the door to her husband he didn't look himself. His glasses were broken and his face seemed distorted. His hair was matted and bedraggled, and his denim overalls were dirty. He wasn't the Bert that she knew. She put on the kettle and ran a bath.

He appeared very dazed, and could only tell her that he wanted to go to Dover. While he had a wash she folded his clothes. They were heavy: the pockets were filled with earth.

Bert had travelled with another soldier and it was from him that Jane learned what had happened. 'He'd kept from me that he was doing bomb disposal, and it wasn't until then that I knew that he was doing the very dangerous job that he was doing.' All he had told his wife was that he was doing demolition work.

He returned to barracks the next day. There he had an interview with an officer, who told him that the best thing for it was to get straight back into bomb disposal, as if he had fallen off a bicycle. But something was wrong with Bert: 'I kept shaking and things weren't right with me . . . I kept wanting to get away.' He was confined to barracks for a week after being caught trying to catch a train at Victoria Station, and then sent under escort to the psychiatric ward of Sutton Hospital.

Woolhouse was kept under supervision for two months at Sutton. The ward 'was a bloody madhouse' and while Bert knew that he was not mad, he 'was terrified in case I had to go back into the Army because I realised that I wasn't the same as I used to be'. In Jane's words, his nerve had been broken.

He was treated by a woman psychiatrist, Nellie Craske. There were pills and graphs, and she talked to him about his childhood, and his sex life, which he found surprising. When Jane was allowed her weekly Sunday visit Dr Craske asked her about the same taboo subject. The doctor explained that Bert did not want to go back to the Forces, and that she wanted to know what sort of marriage he might be returning to.

What Craske did not mention was the controversial nature of the treatment that she had prescribed for some of her patients. With the psychiatrist William Sargant, she was studying experimental ways of treating 'war neuroses' – what had once been called shell-shock. Men who had broken down under the stress of war were being given insulin to induce in them a state of semi-coma as a way of calming their 'anxious, depressive or hysterical symptoms'. Sargant would go on to conduct secret trials of mind-altering drugs on soldiers at Porton Down.

Bert had had the good sense only to pretend to take the pills that he was given by Craske, and was perhaps fortunate to be discharged from the Army by a medical board at the end of 1940. Two other sappers injured at Dagenham were given the same dispensation. Bert's ordeal was not over, however. Jane felt that he had been traumatised by seeing his friends killed, and though talking about it helped, his nerves remained raw. The first time that they went to the cinema after he had left hospital the siren sounded and he became highly agitated. When she asked what the matter was, he replied: 'I can't help it.' She knew no one else who had been through the same experience, nor was she offered any advice or counselling. They had to get through it on their own.

Jane was proud of the work he had been doing but, she recalled, 'it was very difficult at first, because the least little thing upset him . . . I had to be very careful with how we went about things, otherwise he'd blow his top.' Woolhouse had been profoundly changed by his experience and it was years before he no longer had nightmares in which he was torn from sleep by a bright flash. He wrote once to Jackie Lewis's mother, but though she got in touch several times to ask to meet him, Bert never followed it up. The idea of his being alive, while her son was dead, was simply more than he could bear.

12

STILL FALLS THE RAIN

While Woolhouse had been waiting for Ash to arrive at the Dagenham site, another car had drawn up beside him. 'All of a sudden, a fellow comes up with a black Austin 7, a civilian. He was a huge man, about six foot four, and he had an Anthony Eden on and a long black coat. And I said: "I'm sorry mate, you can't come here, you've got to be three hundred yards away." And he said: "I'm the Earl of Suffolk – I've been sent by Bomb Disposal Headquarters."'

Following Suffolk's triumphant return from France, Gough had not immediately been able to find a niche for Wild Jack's talents at the MOS. Suffolk had briefed Special Branch on the political persuasions of the scientists that he had rescued – one professor was 'extremely violently Bolshevistic, while his wife is actually a member of the Nazi Party' – and then had been given a liaison post with the Free French forces. Gough himself was occupied by a highly distracting battle within his own department.

In late summer he had voiced his concerns to the Ordnance Board about their having diverted his Research Department (RD) at Woolwich on to their own lines of enquiry, in effect usurping Gough's BD functions. Their response was that as the

Unexploded Bomb Committee met only once a month, they had nowhere else to turn for urgent advice on fuzes.

With hair-splitting sophistry, Gough pointed out that the UXBC's remit was only to consider problems about bombs, not to collect or disseminate information about fuzes. Given the death of Eric Moxey, he was also perturbed by pressure that had recently been put on Merriman by the service ministries to recover fuze specimens for them. These demands were only likely to increase now that RE units had been forbidden to attempt such operations following the discovery of the Zus and the anti-disturbance (50).

Gough suggested a compromise: the RD would take on the role of investigating fuzes, while the UXBC would devise general methods of dealing with bombs. Not surprisingly, in mid-September the Chief Superintendent at Woolwich, Brigadier J. L. P. Macnair, rejected this proposal, which would saddle RD with a dangerous task yet leave the reins in Gough's hands. His decision had two important consequences for bomb disposal research.

The first was the formation by Gough at the start of October 1940 of a Research Sub-Committee (RSC) of the UXBC. While the latter was to still concern itself with the broader issues of BD policy, the new body would be a weekly conference of scientists from the key ministries and laboratories that could react more swiftly to technical problems requiring rapid solutions. It would be given its objectives by the weekly tri-service meeting of BD 'users' – Gough's term – and in effect would become the dynamo driving all research and development in the fuze war. The RSC's members were to include Andrade, Bernal and Ferguson from Woolwich, as well as Gough himself.

The other consequence of the *froideur* with Woolwich sprang from the first, in the creation of an experimental unit that could supply the RSC and its boffins with examples of fuzes, and which would try out new methods in the field before equipment was put into production. Gough told General Taylor that the lack of such an organisation was 'a very serious gap in our present scheme of things', and by early October he had found a suitable leader.

Solly Zuckerman, who was still working with Desmond Bernal on effects of blast, was among the first to discover Suffolk's new role. They both attended the wedding of Paul Libessart, the artillery expert whom they had known in Paris before he escaped with Suffolk, and Zuckerman noticed that although it was a warm day Jack was wearing a scarf and overcoat. 'At one moment when we had to stand up, Joan [Zuckerman's wife], who was seated between us, teased him about his unseasonal clothes, to which he reacted by unbuttoning his coat, revealing a bare chest covered with a blacksmith's leather apron – the uniform he wore when dealing with unexploded bombs.'

Suffolk had first developed an interest after persuading Gough to take him along to see a bomb being defuzed at Deptford West Power Station. This may perhaps have been done by Merriman, the Experimental Officer in Gough's Directorate of Scientific Research, who was soon due to be posted (like Davies) to Cairo. Gough was therefore looking for a replacement. He needed someone who understood science, who had military experience, whom he trusted and who would voluntary risk their life. It seems to have been Suffolk who suggested himself for the job.

By 7 October, the day that Woolhouse met Suffolk, the earl had already set up his base in Richmond Park. The four square miles of park had been closed to the public for the duration of the war, and the wilderness that had been the preserve of the red and fallow deer was given over to the khaki and camouflage of military use.

Convalescing troops were housed in White Lodge, the home of the King when he had been Duke of York, while several Army units were quartered in other buildings and in the Park itself. Among these was Phantom, the reconnaissance regiment, which used it as a training centre. Later in the war, from their mess situated on a high ridge above the Thames, its officers could see – at what appeared to be almost eye level – V–1 missiles hurtling the last few miles towards London.

More remote areas of the Park were used for secret research.

Curious creations by Robert Watson-Watt, the pioneer of radar, sprouted between clumps of woodland. Another frequent visitor was Desmond Bernal, who had half-sized streets and houses built to test his theories on the way that blast was channelled by walls and alleys. He also experimented on air-raid shelters there, concluding that the roof should be reinforced with concrete if it was to resist penetration by a bomb.

On one occasion he was accompanied by Peter Danckwerts, a young naval BD officer who was later to win the GC for defuzing parachute mines. What impressed Danckwerts more than Bernal's ideas on the fallibility of shelters was the 'perceptible interval between the explosion and the time at which Bernal ducked down' in to their trench. Bernal explained that he knew the velocity of bomb fragments and could calculate their arrival time in his head.

The Park also served as the main proving ground for experimental work on UXBs. Autumn had arrived, and the gaunt frames of oak and elm trees were soon casting shadows on a formidable arsenal of steel and explosive assembled from all over London. A photograph taken at the time shows Suffolk, wrapped up against the chill in a scarf and Homburg, and with his cigarette holder clamped between his teeth, seated on a huge parachute mine. A more conventional man might have had himself depicted with it as a trophy, his foot on another head to be mounted on the wall, but Suffolk's relaxed demeanour is that of someone on a park bench.

Lord Clanwilliam, who had first steered Suffolk towards the MOS, met his fellow nobleman one morning when he was on his way to Richmond to deal with some unexploded devices. 'I asked him if there were many there, and he said, "Yes, twenty or thirty." I saw him in the evening and said how glad I was to see him alive. He might just have come back from a day's shooting.'

At his own expense, Suffolk had kitted out a Pickfords removals van as a mobile laboratory, emblazoned with the letters BD. To the faithful Beryl Morden he had also added a new member to his team. This was Fred Hards, the van's usual driver, who was soon to prove his worth as a handyman and 'quite fearless' improviser.

Hards's wife Elizabeth, whom he called Nin, did not at first believe his stories that he was working with an earl on unexploded bombs. He had always been good at pulling her leg. Then one evening, when she was bathing the youngest of their children at home in Penge, she heard her husband coming in the door. With him was a tall stranger, tripping over the long scarf that he wore. 'Never mind, Fredders,' he said, when Nin became flustered by his presence and her wet arms, 'let's have a cup of tea.'

Suffolk became a regular visitor, often accompanying Hards home with a pocketful of fuzes or bomb parts. One night they brought Chinese food and he showed them how to eat with chopsticks. He liked to be with the children as Fred played games with them, pulling them up through his legs while they were upside down. Suffolk even went with Hards to watch Crystal Palace play. The two men were about the same age, and Hards's company perhaps offered Suffolk something of what he had experienced in the outback, the chance to see and be a part of a normal life. He even offered Hards a job working for him in Australia after the war. 'It'll be good for the kids,' Fred promised his wife, and asked her not to worry about his work in the meantime, pointing out that it had spared him being called up by the Army and taken away from them.

The first scheme that Gough asked Suffolk to devise was a method for burning out the filling from bombs that could not be moved or steamed out, or where corrosion prevented the removal of the base plate by hand. There had been some earlier trials with thermite, and now he explored this technique more systematically. A beaked crucible containing about six pounds of thermite was placed about three or four inches above the nose weld of the bomb's casing. When the charge was fired a molten stream of iron cut a half-inch hole in it 'within approximately four or five seconds'.

In cases where the fuze had already been extracted, Suffolk

discovered that once the resulting fire had been quenched, any powdered explosive could quickly be washed out with a high-pressure hose. This would become a useful and far less laborious alternative to steaming out defuzed bombs. If a bomb was still fuzed, however, the thermite provoked an explosion as the gaine overheated. This led many BD units to regard thermite as a crude and unreliable solution. Nonetheless, the method did burn away up to two-thirds of the main filling before it detonated, and until the spring of 1941 it was used for the rapid disposal of UXBs where explosion was permissible. Certainly, by late October Gough was writing to Taylor that thermite 'as a reliable means of opening the bomb case . . . was proven'.

Furthermore, Gough suggested to Taylor that he now lend Suffolk a section of Royal Engineers to help him with his tests. Given the pressing need for effective means of bomb disposal, there was no time for research to be progressed step by step from the laboratory to the finished product. For Gough there was 'no other practicable course' except to 'proceed from early development direct to tests on *filled* enemy bombs'. The troops could help Suffolk with his trials and be trained by him in any new methods.

Gough was at pains to point out that it was not his intention that, as civilians and volunteers, Suffolk, Morden and Hards should form a permanent part of a bomb squad. Yet as he himself acknowledged, 'after several inexplicable fatal accidents to BD' sections the supply of fuzes to the MOS had dried up. 'But fuzes had to be made available or pressing problems would remain unsolved, and the whole organization of bomb disposal receive a setback.' Gradually, therefore, Suffolk began to offer 'to obtain the minimum supply of fuzes', especially new models, so as to allow the development of 'antidote methods of handling'.

From Bury St Edmunds he returned with the first 250-kilogram anti-personnel bomb to be recovered. Other exploits included the defuzing in early December on the Barnet by-pass of a five-hundred-pound (17)/(50) combination: both fuzes, including the Zus of the ticking (17), were recovered intact.

Although fully aware of the dangers, Morden and Hards insisted on remaining with Suffolk while he worked on bombs. Hards would hand Suffolk the tools needed, while Morden sat calmly beside him, taking dictation as Suffolk called out each move that he made.

As autumn shaded into winter they became his left and right hands, assisting him and MOS scientists in two series of experiments. The first was the use of shaped plastic explosive charges to cut open live bombs, a technique described by one of the earl's colleagues as 'only about 80 per cent safe'. The second, suggested by the indefatigable dowser Merrylees, was the immunisation of clockwork fuzes by shooting them. A carbine, carried in Suffolk's van, was specially modified with a muzzle attachment so that a .303 bullet could be fired into the face of a fuze from various angles. Trials with dummy brass clocks were carried out, and showed that the mechanism could be shattered in this fashion, but the chance that the bullet would hit something sensitive, such as the fuze cap or the gaine, eventually proved to be as high as one in two and a half. The charmed life that Suffolk, Morden and Hards seemed to lead amid all these dangers soon gained them a new nickname from the soldiers who watched them at work: The Holy Trinity.

As Britain continued to be pounded from the air, dozens of experts were engaged in BD research. Progress was made to the rhythm of a scientific foxtrot – two steps forward, and often one step back rather than sideways. In its first report, in early November, the Research Sub-Committee identified the ten types of fuzes so far found. Of these, the (17) and (50) in particular still eluded any solution.

Against the latter, three techniques were being tried by different establishments. Woolwich was looking at the degree of motion necessary to detonate it, while the Cavendish Laboratory at Cambridge was bombarding its capacitors with radon and neutrons to disperse their energy. Meanwhile, the Government

Laboratory, at Aldwych, had been exploring the possibility of introducing oil foam into the fuze to isolate the trembler contact, but the discovery of an updated model in which the trembler was wholly enclosed had put a stop to this. Now hopes were pinned on a procedure that involved blowing steam through the fuze in order to conduct away its charge, the advantage being that this could be done without agitating its hypersensitive switches.

Similarly creative but unsuccessful treatments had been tried against the (17)'s clock. Attempts to drill into it had been abandoned as too liable to disturb the steel timing disc and so release the striker. Hydrogen chloride had been pumped into it, but without effect. Viscous engine oil did stop its movement, but had proved too difficult to inject. Efforts were now concentrated on the magnetic clock-stopper.

It was well known that magnets interfered with the movement of watch parts, and therefore the Germans had taken care to fit their fuze clocks with a non-magnetic balance wheel and hairsprings. Woolwich thought that nothing could be done, but by October the Royal Aircraft Establishment had found a weak spot: the bearings of the third and fourth wheel of the driving mechanism had steel spindles and brushes. Working with GEC, they showed that if sufficient friction was generated in these by the pull of a magnet, this would counter the motion transmitted to the wheels by the mainspring.

Producing a magnetic field in a hollow container such as a bomb was not easy, and much intensive study of this had had to be done. Moreover, the tests demonstrated that individual clocks had their foibles (one had been stopped by a desperate BD officer with a jarring blow from a sledgehammer) and thus a generous safety margin and a very high current would be needed.

A succession of prototypes was built, all based on the principle of a vulcanised coil of wire as thick as a man's arm placed around the axis of the fuze. Through it was passed a surge of 200 amps for about two seconds, and the resulting field was strong enough to bring the clock to a stop. GEC's Mk II version, which was first delivered to troops in November, was named for the process that underpinned it: the Kramer Immunising Technique, or KIM.

The clock-stopper was not without its drawbacks. The first model drew on fourteen 12-volt batteries, carried in three large cases, and the magnet alone weighed as much as a heavy man. Later types were less bulky, and their glass winding insulation allowed the coils to be used at a higher temperature and with less chance of them burning out.

Once a fuze had been stopped, its mechanism was usually not strong enough to start itself again without a jolt. Yet there were always exceptions and so it was vital to know if a fuze began ticking while work was being done on it.

To that end, an electrical version of the medical stethoscope was developed in the late autumn. Attached to the bomb by a magnet, it had a crystal microphone, a headset and an amplifier that allowed any ticking to be easily heard. Experienced BD squads would from time to time place a wristwatch on the UXB, just to check that their comrade had not eased off the headphones to give his ears a rest. Such lapses could be lethal.

When used correctly and used together, as they were starting to be from early December, the stethoscope and KIM began to even up the odds between BD and the delayed-action fuze. Winter had come, but spring could not be far behind.

The change in season had also signalled a change in strategy by the Luftwaffe. By mid-October it was evident that no decisive blow had been landed against the RAF, and thus that no invasion could be mounted this year. German Army Intelligence had come to believe that the Luftwaffe's assessment of the damage that it was doing was too optimistic, and suggested a change of tack. Believing that much of Britain's capital, as well as a high propor-tion of its production capacity, was based in the Midlands, it suggested widening the front to target the provincial arms towns.

Aside from Liverpool, hitherto only London had been bat-tered consistently. Even in November it was to be bombed every night but three, with twice as many parachute mines – 1215 – falling as in any other month. In the first few weeks of raiding, three boroughs had been struck by at least a hundred bombs per square mile every evening, and by early December more than forty thousand of them had landed on the capital. Although attacks were to become less frequent, in proportion to its population London had by then been hit eight times harder than any other British city.

When the targets became Coventry, Birmingham, Manchester and Hull – let alone the likes of Nuneaton, Scarborough or Weston-super-Mare – the effects of bombing were far more devastating than they had been in London. Its sheer size meant that even if one part was badly damaged, such as the Square Mile, others were still able to provide essential services. If the clothes shop in the borough's high street had been hit, there were alternatives within walking distance. There would of course be some social dislocation in the first days after a heavy raid. Where houses had been damaged, people might have to stay somewhere else for a few weeks and, until communications were restored, for a while shopping for food became an adventure. Yet the heart of London – Piccadilly, Oxford Street, Theatreland – always remained recognisable, and life went on.

In smaller places, the effect of just one heavy raid could be catastrophic. Indeed, after the obliteration of the centre of Coventry on 14 November a new word was coined by the Germans to describe the near-eradication of a town's identity – '*Coventrieren*', to 'Coventrate'. Outside of London, even, in large cities such as Manchester, there were few important civic symbols or much shopping beyond the historic centre, and accordingly the effect of their destruction tended to be more pronounced than it was in the capital.

In Southampton the shops on the long high street burned to the ground. In Liverpool the grand offices of the shipping companies were destroyed. In Manchester half of Piccadilly was laid open to the sky. Emerging into the morning light after an attack, people would struggle to get their bearings amid what was suddenly a mass of ruins. The population and local government in the provinces were less accustomed to bombing than were Londoners, and in some of the smaller towns the morale was much worse hit. In Coventry itself, it came close to collapse.

'There were more open signs of hysteria, terror, neurosis, observed in one evening than during the whole of the past two months together in all areas,' concluded a report on the aftermath of the great raid. 'Women were seen to cry, to scream, to tremble all over, to faint in the street, to attack a fireman and so on . . . There were several cases of suppressed panic as darkness approached. In two cases, people were seen fighting to get on to cars which they thought would take them out into the country, though in fact, as the drivers insisted, the cars were just going up the road to the garage.'

Coventry did sustain exceptional damage. Some four or five hundred shops had been destroyed or were forced to close, and rationing had to be suspended as feeding arrangements broke down. Thousands of people fled the city on the following evenings in case the bombers returned. They did not, and Coventry recovered, and was strong enough to resist two more raids in April 1941. Yet even after its houses and factories had been repaired and reopened, its psyche still bore the imprint of the destruction in a single night of so much that had stood for centuries.

The damage to towns such as Coventry was exacerbated by the Germans' use, from November, of a larger weight of incendiaries, with raids regularly launched on Sunday nights when there were fewer workers about to put out fires. The performance of the provincial fire brigades did not help matters. Independent from one another, they had no standardised commands or equipment, so that when they arrived to help their neighbours their hoses often could not be connected to hydrants. Many also charged for their aid, leading to property being left to burn in order to save money. A national fire service was not created by the Home

Secretary, Herbert Morrison, until May 1941, while he only made fire watching compulsory after Manchester had burned at Christmas 1940.

The switch in focus to the regions put more strain on the BD units there, some of which had yet to be severely tested. The medium-sized metal industries of Birmingham became one frequent target, and two thousand people were killed in five attacks in the autumn and early winter. On 28 October, for instance, Second Lieutenant Ralph Lee of 9 BD Company was called to an aircraft factory at the city's Leebank Works. A 250-kilogram UXB had torn through the building and come to rest underneath some machinery. With the help of the factory's auxiliary BD unit, led by Reg Cooke, a shaft was sunk beneath this. The digging continued for two days, and when the bomb was finally reached it was found to be armed with a fuze that had been ticking all the while.

With two other men, Lee attempted to extract the (17), but without success. Seizing a crowbar he next managed to lever out the electric portion of the fuze. The clock, however, remained within and still working. In desperation, Lee filled first the fuze pocket and then the entire shaft with water. This did stop the clock, and the bomb was removed the following day. Both Lee and Cooke were subsequently awarded the George Medal.

Coventry was bombed several times before the attack of 14 November. One raid in mid-October left a 250-kilogram bomb embedded in the Triumph motorcycle works in the centre of the city. Two factories that manufactured war materiel had to cease production, and more than a thousand workers had to be evacuated, as did nearby residents. It was essential that the bomb be neutralised as soon as possible. The task fell to Second Lieutenant Sandy Campbell of 9 BD.

Although a junior officer, Campbell was at forty-two another of those middle-aged professionals who had been posted to Bomb Disposal; in peacetime, he had been an electrical engineer on the Isle of Bute. The UXB, he discovered after two days of digging, had a (17) fuze, no doubt protected by a Zus, and so he would be unable to take it out to make the bomb safe. However, it was not ticking, so Campbell decided to take a chance. He had the missile loaded on to a lorry and – knowing exactly the risks that he was running – then lay down beside the bomb so that he could hear the fuze and warn the driver if it did start to function. They drove slowly away from the site for about a mile, and then blew up the UXB.

The following day Campbell had to deal with another 250-kilogram bomb which had fallen two days earlier on a housing estate in the city. With him this time was Serjeant Michael Gibson, a collier from Chopwell in Tyne and Wear. Gibson had fourteen years' service with the Durham Light Infantry (Territorials) behind him and a month earlier had shown his mettle when confronted at a factory with a UXB that began to make a loud hissing noise. Although another bomb of the same type. had exploded near by, he continued to work on it, having told the rest of his section to take cover. Eventually he managed to defuze the bomb and save the workshop.

The UXB at the housing estate had a straightforward (25) impact fuze, and although this proved impossible to remove it was discharged normally and the bomb was taken by road to Whitley Common. As it was being unloaded it detonated, killing Campbell, Gibson, four sappers and the driver. It was thought afterwards that the fuze had somehow retained enough energy for contact to be made while it was being shifted.

The explosion was seen from a distance by Michael Gibson's two young sons, who were unaware of its origin. Gibson himself was unaware that only a few days earlier he had been approved for the GC following his bravery at the factory. Sandy Campbell also received a posthumous GC, for his actions at the Triumph works.

The codename *Mondscheinsonate* (Moonlight Sonata) was given by the Germans to the four-hundred-aircraft raid launched against Coventry four weeks later, timed to coincide with the full moon.

A month on it was to be *Tiegel* ('Crucible'), and the turn of Sheffield.

On the night of 12 December Lieutenant John Hudson was woken from his sleep by a distant but heavy rumbling. He got up and went to the window. Outside, milky moonlight poured down on Halifax, but the dark line of the horizon was backlit by bright flashes. Thirty miles to the south, Sheffield was being pulverised. 'I thought, That's going to be work for us. And sure enough, first thing in the morning the phone went.'

It was his Commanding Officer, ordering him to take his section – No. 59 – to Sheffield, where more than a hundred UXBs had already been reported. Hudson had had just three weeks' experience with BD since his training, and he had yet to defuze a bomb. More pressingly, he now had to work out how he was going to get his men to their destination. 'My transport was one baker's van and one lorry, neither of them in very serviceable condition . . . One of them had gone off to collect the rations and the other wouldn't start.

'So I rang the local bus company and told them I had to take a party with all haste into Sheffield. They said they were glad it was me not them, did I know there had been a raid, that sort of chit-chat. So they sent a bus, and we all got in and went off to Sheffield. I got a hell of a rocket, because the bill went to my CO, and I hadn't got any authority to hire a bus. I thought it was rather resourceful.' It was a quality that, over the next four years, was to prove invaluable to BD.

John Hudson had been in the war since its start, but not as an officer. Nor did he come of traditional ruling stock. His grandfather was a gardener from Macclesfield, born of a family where the only spirits permitted were of the holy kind. John's father Arthur ran the post office at Chapel-en-le-Frith in Derbyshire, having learned to work the telegraph while in East Africa with the Royal Engineers during the Great War. His malaria seldom bothered him, but there was a brooding restlessness to him, a frustrated intelligence that overshadowed his wife Bertha and John's sister Molly. It was John who would be the vessel on which he loaded his own hopes for the future.

Though he was a clever boy, with an aptitude for mathematics and science, John was not an ambitious one. In 1926, aged sixteen, he left his secondary school with no clear idea of what he wanted to do. The only path to which he had been drawn – studying physics – was closed off by a teacher who had told him that shortly everything worth knowing about it would be known. His father, however, did have a plan: that John should work for him at the nursery he was intending to open, and so he was enrolled on a one-year diploma course in horticulture at an agricultural college outside Nottingham. When, the following summer, he came top of the Royal Horticultural Society's exams, it was suggested that he go on to take a degree in the subject. So it was that slowly he began to find his own way, and to leave behind a world beyond which he had not thought to stray.

Much of the impulse to do this came from someone he had met, Gretta Heath, the sister of one of his lecturers. The only girl among six boys, Gretta came from a more middle-class background, but her mother had struggled to bring the children up after their father, who managed a farm, had died when they were young. Four of her brothers migrated, but Gretta's mother had other ambitions for her. These perhaps did not encompass the quiet loner that Gretta had met, but in John she found contentment, while from her he drew resolve. By 1936 he had taken a job as a horticultural instructor for East Sussex County Council, and that year they married.

By then, John later recollected, it 'looked like there might be trouble on the Continent'. As a way of encouraging men to become Territorials, his employer announced that in the event of war the wives of members of staff who were called up would continue to receive their husbands' full pay if they had been in the TA. 'So of course we all joined up.' The Army appeared to think that a second world war would be much like the first, and even when Hudson found himself sailing to France in March 1940 as part of the BEF he had learned little in eighteen months beyond foot drill and how to dig a trench.

Following his father for once, he had joined the RE Territorials, and had been assigned to a Field Park Company, which looked after stores and heavy equipment for other units. He had applied for a commission, to which he was entitled since he had a degree, but had been told that there were no vacancies. Hudson therefore began his military career as an ordinary sapper.

By late May the BEF was conducting a disorderly retreat. Having been told to abandon his vehicle, the roads being jammed with lorries, Hudson soon became separated from his comrades. Cadging lifts as he could, walking when he could not, he eventually arrived at a small coastal town named Dunkirk. Spread out before him was what remained of the British Army in France, now in a state of chaos. 'It was . . . astonishing in a public street to see parties of soldiers . . . walking in groups like parties of hikers out for a ramble. Nor had we before seen British soldiers in public with stubbly, long-unshaven chins.'

For several days he waited on the dunes for rescue. 'One chap, with an unshakable faith in the Guards, then defending the perimeter, said that the tighter we were squeezed the harder we should counter attack ("See what happens to a spring when you squeeze it"), overlooking the fact that most of the British Army was on the beach with no equipment, getting its feet wet, and peering out to sea for signs of shipping.'

Long queues were formed for those boats that did appear and, showing their gift for improvisation, the soldiers made piers to them from planks placed atop a line of lorries that had been driven into the water. 'But as the tide came in, those at the front gradually found that they were in deep water, standing in deep water.' Attempts to make those behind them move back were fruitless, and Hudson saw at least one man fall in and drown, pulled down by the weight of his greatcoat. The sight of bodies

lying in the sun by the edge of the water, and his shock at realis-
ing that they were dead and not asleep, was to remain with him all
his life.

Hudson was among the fortunate ones who made it back to
England. He was amazed by the reception that greeted him: 'We
left France as a defeated army in full retreat. We were astonished
to discover when we got to Folkestone that we were welcomed as
heroes. Pretty dishevelled, too, we were for heroes.' Taking advantage of the confusion at the station, he attached himself to a
division bound for Yeovil, where Gretta and their two young
boys were staying with her mother.

'You see by now we weren't really a disciplined group any
more,' he admitted. 'We arrived in Yeovil in the middle of the
night . . . I slipped off behind a hedge to have a pee, let the
column march on – nobody knew who was there or not – then
when they were out of sight I walked home and knocked on the
door. My wife was astonished – she'd been worried stiff listening
to all the Dunkirk news – to see me there.'

After a day or two Hudson rejoined his company, only to be
surprised when he was first promoted to sergeant and then shortly
afterwards informed that he was being commissioned into Bomb
Disposal. At the start of November he was given command of 59
Section, one of the four that made up 14 BD Company. This had
its headquarters in Leeds and was led by twenty-five-year-old
Raymond Bingham. Such was the rate at which BD was growing
that while Bingham had been a second lieutenant in mid-
September, by early December he would be a major. Hudson,
meanwhile, was sent to train at Melksham before joining the section at their Halifax billet a month before Christmas.

There had been arsenals in Sheffield for almost a century, since
Henry Bessemer had set up his steelworks in 1859 at Attercliffe.
More than half a million people now lived in the city, many of
them employed in the armaments factories and heavy industry
strung out beside the River Don. Guided to their objective by the

newly developed X-Gerät system of radio beams, German pathfinder aircraft arrived over Sheffield at half past seven on the evening of 12 December. The first of eleven thousand incendiaries began to rain down into the cold night air, raising a halo of fire that would draw in like a beacon the oncoming fleet of almost three hundred aircraft. The raid continued until four o'clock the following morning.

Its targets included the Vickers foundry that housed the fifteen-ton drop hammer used to stamp the crankshafts for the Rolls-Royce Merlin engines that powered the Spitfire and the Lancaster bomber. It was the only one of its kind in Britain. At another target nearby, the Hecla Works, were made all of the country's eighteen-inch armour-piercing shells.

So fiercely did the inferno burn in the city centre that tramcars were welded to their tracks by the heat, blocking fire engines from getting to the main seats of destruction. For nine hours cinemas, shops and churches were shattered and scorched by an avalanche of high explosive. At half past nine the west end of the cathedral was demolished. As the blaze spread among the commercial buildings in the centre, those who had taken shelter in basements there were forced to flee into the open. More than two thousand people were moved to safety between a double line of fires along one main street alone. Communication between the emergency services broke down entirely.

Just before eleven, a bomb demolished the C&A building in Fitzalan Square. Then, an hour later, the Marples Hotel opposite took a direct hit, bringing seven floors crashing down on some eighty people sheltering in the cellar. Seven of them were brought out alive, some after having managed to dig an air vent with their hands; only fourteen of the corpses could be identified.

Hudson and his thirty men drove into a city wrapped in fog and reeling from shock. Familiar landmarks had vanished. Entire districts had been burned to the ground or were choked with rubble. A map would have been useless, and Hudson was given as a guide a local policeman from the West Bar station, Sergeant

Buckler. Together with his own Serjeant Battersby, they began to reconnoitre the most-urgent sounding of the 130 UXBs assigned to Hudson's section in the south of Sheffield. Two bombs went off near by, in sectors given to their sister sections, but three-quarters of the reports Hudson investigated turned out to be false alarms.

'A lot of the wardens were old people,' Hudson later reflected, 'who were scared, no doubt. It was quite clear often when you got to a site that it was not a UXB, but a bomb that had gone off. Or two chaps had reported the same bomb from different directions.' A large hole in a roof suggested the presence of something, but it turned out to have been made by an Austin 7, which was now perched in the attic; it had been blown straight up into the air and on to the house by the blast from a landmine.

Nothing could be done about the thirty remaining genuine UXBs for another four days, in case they turned out to be time fuzes: 'Experience over the whole country had shown that bombs were most unlikely to go off after ninety-six hours until disposers were actually dealing with the bomb fuzes, when people might be asked to leave the area again for a short time if there was any serious risk of an explosion.' Hudson sent the section back to Halifax, but two of the bombs were rated Category A and needed to be dealt with by him immediately. By the time he and Buckler reached the first it had detonated. The second was near a telephone exchange.

'It had fallen into a cottage two or three doors away. There was a hole in the roof. . . . and there it was sitting on the cellar floor.' It was a fifty-kilogram bomb, the first of any kind that Hudson had had to defuze. He knew that its fuze pocket was too small to conceal a Zus, 'so I took my key out, unscrewed the locking ring, took out the fuze, unscrewed the gaine, put it in my pocket, put the fuze in my other pocket, and then the sergeant put it on his shoulder and we walked away with it'. They eventually dumped the bomb in a hole in a factory yard where it could do no damage.

During the next two weeks the section learned on the job. Like

Hudson, with whom they had an easy relationship, none were regular soldiers. Now they became skilled with the basic tools that were still all they had – picks and shovels and the thin steel probes with which they sought to trace the track of a buried UXB from its entry hole. Yet Sheffield was built not on soft London clay but on limestone and millstone grit, through which a speeding bomb would jink and ricochet. There was usually little that could be done except to dig and to improvise. On one occasion, Hudson's section used a piano as an anchorage for a hoist so that a bomb could be lifted out of a deep hole.

Their task was made more difficult still when the bombers returned on the night of 15 December. By the end of the two raids more than six hundred people had been killed and fifteen hundred others wounded. Some eighty thousand houses had been damaged, and forty thousand of their inhabitants made homeless. The losses caused to Sheffield's industry, however, were slight.

Exposed to more UXBs than many BD officers outside London had seen, Hudson quickly became an expert in the different type of fuzes. On 19 December, for instance, he reconnoitred nineteen and dealt with six. All of these were impact fuzes, but he also successfully navigated a (17)/(50) combination in which the Zus had failed, and helped the Navy to burn out a parachute mine that had landed on the railway station.

He came to appreciate that even the final disposal of a bomb was not without its hazards. The section blew up one in the middle of a field, only to discover that in so doing they had cut the water main that led from one of the Lake District reservoirs to the city. Potentially more serious was Hudson's attempt to finish off a UXB that had a time fuze which had stuck after being partially extracted. It was taken out on to Ringinglow Moor above Sheffield.

'We rolled the UXB a few yards from the road,' he later wrote, 'and my chaps withdrew, whilst I left my car alongside with its engine running to make a quick getaway after setting a gun cotton charge on top of the bomb. After lighting the touch I got back in the car, but must have been over-anxious because the gear lever broke off in my hand, leaving the engine in neutral.

'There was no time to run, so I lay flat behind a low wall at the roadside; within seconds there was a colossal bang and a shower of wet black peat completely covered both the car and me. There was no hurt, except to my pride, but I doubt if many have been as close as that when a large bomb exploded and lived to tell the tale.'

All these experiences contributed to a report that he decided to write after lunch on Christmas Day. This had been 'celebrated in the time-honoured fashion by all Sections', and Hudson recalled that he might have had a beer or two before sitting down at his desk. 'I wasn't asked to, it wasn't required, but for some reason I took pen to paper and wrote about what we'd done.' The letter, which was addressed to his CO, gave a frank assessment of what the work of a BD section in the field was like, and how it was being hampered by shortcomings in equipment and supplies.

Hudson's perception was that this was largely due to the fact that those at Headquarters had no practical knowledge of the labour involved. He, however, had seen matters from both sides, as an officer and in the ranks. 'Salmon and meat-paste sandwiches,' he pointed out, 'are not the ideal mid-day fare for men working day after day on heavy excavation work.'

What use was a cooker in a bombed city where the water and gas had been cut, and no fire could be made after dark because of the blackout? Why not make one officer responsible for rations and fuel when several sections were at work in an area, as officers rarely had time to see to this otherwise? Should not section boundaries in a town coincide with police divisional ones to avoid duplication of UXB reports? And would not derrick lorries – as issued to RE Field Park Companies – be a more serviceable mode of transport than the worn-out Austin 7 that currently constituted half his section's motor pool?

A few days later Hudson was summoned to see his commanding officer. He now realised that he might have been a little incautious in giving voice to his criticisms. And indeed his CO did have news that he was to be posted away – but to London. He had been so impressed by the report that he had sent it on to Bomb Disposal Headquarters, where it had created quite a stir. As it happened, they were looking for an officer with hands-on knowledge of bombs who could act as a link between them and the sections. Hudson's function would be to ascertain their requirements and to ensure that they were following the latest instructions about fuzes. To this end he would be working directly for General Taylor, and be liaising closely with Dr Gough and his team. It was all a long way from tomato plants and silage.

On his last morning alive, Captain Max Blaney had told his sergeant Len Jeacocke that he was off to deal with 'a sticky one' in Manor Park. By the time that he heard of Blaney's death, and those of the nine others killed with him, Jeacocke was at work himself on a bomb in the garden of a bookmaker at Chingford. He suspected that it might have a ticking (17) fuze. 'I thought: Is it or isn't it?' Like Blaney, he had noted that it was Friday 13th, and so decided to be extra careful.

Having borrowed a doctor's stethoscope, he 'got back into the perishing hole'. Space in the shaft was tight, and he had to reach around the bomb to place the microphone over the fuze. 'From it came an ominous and rhythmic "thump, thump, thump".' Only after he had made a hasty exit, and an electric stethoscope had been brought up to the bomb, did he realise that the noise that he had heard was the rapid beating on the casing of his own heart.

'Lady Luck was shining on me,' wrote Jeacocke about the incident that had claimed Blaney's life the same day – 'Fortunately, he didn't take me down to it with him.'

Three hundred people attended Blaney's requiem Mass early on a cold December morning. Many of those present had to leave straight afterwards to return to work on the hundreds of UXBs still tying up large parts of London. The risks that had been, and were continuing to be, taken by men like Blaney were starting to pay off, aided by a marked decrease in attacks on the metropolis.

This was in part due to the decision to target instead the ports and manufacturing towns, as well as to the onset of severe weather that greatly increased the chances of a flying accident and so reduced the number of sorties mounted by the Luftwaffe. After that of 8 December, which had led to Blaney's death, there were no more big raids before Christmas 1940. The number of UXBs awaiting disposal began to be whittled away, and the festivities offered BD personnel a rare and deserved period of relaxation.

The anti-aircraft guns sounded again on the night of 27 December. It was a clear and frosty evening, remembered Theodora FitzGibbon, who ventured out to buy a packet of cig-arettes in the King's Road, Chelsea. 'The night was bright with a full moon, searchlights and the now customary glow of fires. Large pieces of shrapnel were clearly visible on the pavements as we walked up Glebe Place.'

Ambulances and fire engines were racing towards the Town Hall, and she heard that the Six Bells, a pub that she often fre-quented, had been bombed. She hurried down to it and saw that the top part had toppled into the road. Huge chunks of masonry lay scattered on the ground.

Having helped to clear the rubble she looked inside what was left of the building: 'Curly, the Irish barman, had gone down to the cellar and was found, his rimless spectacles still on his snub nose, but he was stone dead, an unbroken bottle in his hand. Almost half-full pints of beer were standing unspoilt on tables in front of the customers who stared at them with unseeing eyes, for they too were dead from the blast.'

Later that same week, on Sunday 29 December, the bombers returned. They concentrated on one of the least defensible parts of London, the City, which was made more vulnerable by an exceptionally low tide on the Thames, rendering it hard for the Fire Service to draw water from it. For three hours incendiaries fell like pine needles, so that by the middle of the night 1500 fires were blazing between the Tower and the Temple. Six large sectors had to be left to burn themselves out, including a zone half a mile square stretching from Old Street to Cannon Street that would become the widest continuous area destroyed during the war. The main water supply in the financial district failed, pumps sunk into the river became clogged with mud and about a third of the City was gutted. Among the buildings damaged or destroyed were eight Wren churches and the part-medieval Guildhall.

One journalist who was determined to get as close as he could to the flames was Stanley Baron. 'Wherever one looked up the narrow alleys of the City,' he later recalled, 'you saw what looked like red snowstorms. Great showers of sparks were coming from the burning buildings.'

He managed to get inside the entrance of the Guildhall while it was on fire: 'And looking up from the outside the building, I saw the quite extraordinary picture of the silhouette of a fireman on a high water tower and his jet illuminated by the flames of the building.' What struck him most about the night was its remarkable beauty. 'It sounds fantastic but it really was a very beautiful scene. The colours were fantastic. The dome of St Paul's was to be seen against a background of yellow and green and red with great billows of smoke coming from it.'

Those who saw walls of flame a hundred yards long advancing on St Paul's, fanned by a strong wind from the south-west, were convinced that the cathedral was doomed. Fire-watchers on the roof of the *Daily Telegraph* building in Fleet Street observed a stream of incendiaries cascading on to the dome and nave of the church, and as some began to burn like fairy lights in the trees outside American reporters cabled home that St Paul's was on fire.

The best-known of them, Ed Murrow, began to read its funeral rites in his broadcast to the other side of the Atlantic: 'The church that means most to London is gone. St Paul's Cathedral, built by Sir Christopher Wren, her great dome towering above the capital of the Empire, is burning to the ground as I talk to you now.'

Inside the cathedral, the danger had only just been realised. The Watch had been battling to quench with stirrup pumps the incendiaries that had landed on the roof, but it was not until Cannon

Street Fire Station telephoned a warning to them that it was appreciated that the dome was alight. A fire-bomb had lodged in the outer shell and the burning phosphorus was starting to melt the lead of the dome. If the timbers that supported it also caught fire, then they would soon be fanned into a roaring furnace and the great dome would collapse, bringing down much of the cathedral with it.

Walter Matthews, the Dean, had been roused in the crypt from his bed of sandbags which were packed around the effigy of his predecessor, John Donne. He had selected for the patrol of the highest parts of the cathedral 'those with a head for heights and a leaning towards acrobatics', for they were expected if necessary to walk along the slender beams of the dome to put out fires.

Although two teams began crawling towards the incendiary in the dome, there seemed little chance that it could be reached in time. All the Dean could do was pray. And then, prompted by an unseen hand, the bomb toppled outwards and fell onto the Stone Gallery around the dome, where it was easily put out.

'I have to confess that it is uncertain how the bomb came to fall', wrote Matthews. 'At any rate, the Cathedral was saved from one of its most perilous predicaments, whether by human means or by what we call "accident". In either case, we thank God that our great church was spared when the situation looked almost hopeless.' Its salvation allowed Herbert Mason that evening to take the most celebrated photograph of the Blitz, of St Paul's unbowed before the Second Fire of London.

During the winter, the cathedral had come to be London's church. People would look out of their windows every morning to check that it had survived the night's raids. In Bethnal Green, one woman had gone out onto a roof to watch the City burn that Sunday: 'And I've always remembered how I was choked, I think I was crying a little. I could see St Paul's there, and the fire all around, and I just said: "Please God, don't let it go!"'

'I couldn't help it, I felt that if St Paul's had gone, something else would have gone from us. But it stood in defiance, it did. And when the boys were coming back, the firemen said: "It's bad, but, oh, the old church stood it." Lovely, that was.'

Fierce though the raid was, the City was not a residential area, and only 163 people perished in what had been one of the most sustained attacks of the Blitz. If the bombing provoked feeling of helplessness, it also stimulated those of fortitude. And the realisation was growing on both sides that bombing could not by itself destroy national morale – not that that was to prevent the RAF from trying the same strategy against Germany.

Since July and the start of the air war, twenty-three thousand civilians had been killed and eighty thousand had been injured without significantly impairing Britain's will to fight or its ability to resist. Since September and the beginning of the bomb war fifteen thousand UXBs had landed on its cities, and while equipment to deal with them had remained in short supply, they too had failed to hobble the war effort as much as Hitler might have hoped. More and more his mind moved not to a knockout blow directed across the Channel, but on one that jabbed eastwards.

13

ARMS AND THE MAN

It was the experience of being bombed that made up Victor Newcomb's mind for him. Until the family home in North London was hit in the Blitz, he had been determined not to have any part in a war he despised. His registration some months before as a conscientious objector had been approved only on condition that he accepted service with the Non-Combatant Corps (NCC), but he had contemplated absconding when his call-up papers arrived. Then the Luftwaffe had struck.

'During this particular period,' he later recalled of the winter of 1940, 'I had rethought my position both in relation to my family and in relation to the generation of which I felt I was a part. It seemed to me, and I think this was impressed upon me as much as anything by the bombing of London, that my generation was in a mess and my position was alongside my contemporaries.'

Victor was in his mid-twenties, and his convictions had been shaped by the times in which he had come to manhood. The child of working-class Baptist parents, though never particularly religious himself, he had grown up in Stoke Newington. He was bright, having won a scholarship to a grammar school, while his political conscience was awakened by his own experience of the turbulent thirties. His father, a local government officer, had for a time been unemployed during the Depression. Living near the East End, Victor had witnessed the marches of Mosley's Fascists and when civil war had broken out in Spain he had been tempted to join the International Brigade.

Above all, Victor had been influenced by a chance encounter one lunchtime when working as an office boy in the City. Passing Tower Hill, he saw that Donald Soper – the evangelist of pacifism – was preaching. 'There was a great deal of heckling and I felt a great admiration for the first time in my life for a man who could handle this kind of situation.' Victor fell under his sway, and after becoming a member of the Peace Pledge Union initially refused to serve when his friends began to be taken into the Forces. The Blitz, however, was not only to change his stance towards the war; it would also give him the opportunity to undertake the humanitarian work for which he felt fitted.

By November 1940 BD had been stretched almost to breaking point. The sections were coping as well as they could with the deluge of UXBs and in London alone, they had already defuzed more than two thousand. But across the country, almost four thousand remained to be dealt with, and the number was rising with each raid. General Taylor, meanwhile, had already lost almost two hundred men to the bombs, many of them those with the most experience. Every extra pair of hands could make a difference, and when he was offered reinforcements drawn from the NCC he willingly took them despite the misgivings of other officers about their suitability. Nearly four hundred non-combatants were to volunteer for bomb disposal work, contributing more than a dozen sections to BD's strength.

Those in the NCC, which Victor Newcomb joined in the New Year of 1941, were used to being cold-shouldered by fighting troops and civilians alike. Passengers in railway carriages were often curious as to what the initials on their forage cap stood for, not having seen them before. 'And when we explained to them what it meant, most people shut up.' He loathed his training on

Salisbury Plain, where they were just used as labour, and jumped at the chance to join a BD unit.

Those in the NCC had diverse reasons for not wanting to fight. Some were inspired by religious beliefs, others by political motives. Newcomb had fallen in with a like-minded group opposed to war but determined to do what good they could. 'Bomb disposal,' he said on reflection, 'seemed to me to be a way of expressing a humanitarian concern for my fellow citizens.' He was sent to Bedford, which received many of the bombs dropped by aircraft returning from raids on the Midlands.

Although the work was largely confined to digging – an arduous task in fenland, where water rapidly filled any hole – he enjoyed the physical exercise. Less warm was their reception by other BD squads. 'I think to begin with they rather resented our intrusion into this specialist job, and therefore seemed to feel that we were in some way belittling the work they were doing.' RE units were also annoyed when, especially as the bombing eased, they were returned to their usual construction duties and replaced in BD by NCC sections.

'The RE seemed to be heavy on Scotsmen,' remembered Newcomb, 'and I think the Scots among them felt this [resentment] more than anybody else . . . I think they came to the conclusion that we were a lot of old softies anyway. I think this was the image of the conscientious objector painted, I think, by the *Daily Mirror* at the time when it gave occasional reports from tribunals and referred to anaemic spotty-faced incompetents. We were given an image of real cowardice as much as anything else, and associated with a frailness of personality and a frailness of physique, which we definitely were not once we got into bomb disposal.'

Not everyone, it was true, was suited to the demands of BD. The writer Mervyn Peake, who had pacifist leanings, was transferred in the early summer of 1941 from the Pioneers to the BD unit based in Chelsea. The combination of the work and his own temperament provoked a minor nervous breakdown, and having set fire to a wastepaper basket in the Company offices he was hustled out of the Army and allowed to become a war artist, as he had long wanted.

Newcomb, however, was made of rather sterner stuff. Like all Bomb Disposal troops he now wore its red sleeve flash, designed by Queen Mary, which depicted a bomb embroidered in gold threads with royal blue detailing. This identified him to other soldiers more clearly than did the NCC badge, but for two years he was still the subject of occasional insults from fellow sections within BD itself.

'When people wished to ostracise us, the word "conchie", and yellow-belly, and that sort of remark is the obvious one to make,' he confessed, 'and it was made frequently. I found it fairly easy to absorb because the general atmosphere was not so bad.' What was to change it for the better was a football match.

Newcomb and his comrades found themselves sharing a training camp with some regular BD units, and after much baiting the challenge came to prove themselves men on the pitch. 'Everybody was around to witness it,' he recalled vividly, 'because they thought it was going to be the slaughter of the innocents. They really came onto the field with the intent to kick us off it, and they really put on the heavies.'

Things did not go as expected: 'We lost the match. But we lost it in such a way, with such a degree of dignity, and such a degree of skill in the football arts – and a degree of physical resources – that the atmosphere really changed almost overnight.'

Disdain for conchies was not confined to the ranks. Many officers, including Captain Herbert Hunt, were reluctant to employ men widely seen as cowards. Having taken over 33 Section following Ash's death, in December he had become Divisional Officer for BD Group 7, covering East London from Barking to Chigwell. Some non-combatants were added to his sections in March 1941 so as to release sappers for other duties, but Hunt complained on the grounds that they were untrained and that he had a large backlog of UXBs to work through: 'I said much more concerning my private opinion of these types, and that put me into disfavour.'

Hunt contrived to keep them off operations as long as he could, until one day a naval team asked for help with clearing a parachute mine that had fallen into Barking Creek. Dealing with live ordnance there was both a dirty and a hazardous job, involving much rooting around in soft mud and effluent from the nearby Beckton Sewer Works. It was with some relish that Hunt saw 'a chance to lay on a "sticky one"' for the despised conchies.

'It was dangerous for any section,' wrote Hunt, 'but the longer I stood over them and the mine, watching, I slowly realised that these men had real guts, and from hoping that something would happen, bearing in mind that I would be in it as well, I felt a great admiration for them and their efforts. They worked well and there were no grumbles . . . I engaged them on other trying jobs such as sludge farms and sites near rivers which were waterlogged. Little by little the men of the company took them to heart and they became good friends.'

The New Year of 1941 brought a change in German night-raiding strategy. In the previous year there had been relatively few evening attacks on ports and dockyards, but in the twelve months to come only six of the forty night raids outside of London were not aimed at coastal cities. The first of the year fell on Cardiff, where 8 Company's billet at Fitzhammond Embankment was first damaged by high explosive and then wrecked by a mine that landed on Cardiff Arms Park.

Stuart Archer was still recovering from his broken leg, and it was not until a fortnight later, on 16 January, that he was able to report for duty. That evening, Swansea was attacked in strength and Archer was sent down from Cardiff for a week to help out 103 BD Section's CO, Lieutenant W. D. G. Rees. Almost exactly a month later, the Luftwaffe returned.

The 'Three Nights' Blitz', as it became known, began at half past seven on 19 February and continued almost until dawn. The quiet, snow-covered streets of Swansea city centre were ripped asunder by blast and fire. Emerging from the cellars in which they had sought refuge, the inhabitants formed bucket chains that drew their water from the baths of houses still standing. Even so, much of the Edwardian architecture of Dylan Thomas's 'ugly lovely town' was destroyed. By early morning on 20 February, more than a hundred UXBs had been reported.

Three of these were quickly rated Category A. The first had buried itself outside the Baths, and a road-breaker was brought in to unearth it. When it was found to be armed with two (17) fuzes it was hastily left alone, and eventually exploded early the next morning.

The second high-priority UXB was at the Cwm Felin steel works and 103 Section set to work. Some seventy German aircraft returned to the still-smouldering town on the evening of 20 February and the men of the section were forced to take cover under a railway bridge. They were badly shaken when it was straddled by bombs.

Having taken out the impact fuzes in the UXB at the steel works, the section then divided in two. One sub-section was detailed to deal with a new bomb that had landed near the railway line that ran along the seafront. The other, under Lance-Serjeant Tom Henderson, was assigned to the last Category A, in Castle Street, where it was blocking traffic on a major through route. It was soon revealed to be a huge, thousand-kilogram bomb fitted with a motion-sensitive (50). This would normally have been left for several days, but as the site was deemed of high importance this could not be allowed to happen, nor could the UXB be blown up where it lay. The only course of action was to try to steam out the explosive, a highly hazardous operation given that the fuze was sensitive to vibration.

At about four o'clock Rees met Jim Lacey at Castle Street to discuss handing over the keys for a new billet. While he was there, the staff-serjeant supervising the sterilisation, Tom Munford – a regular from Inverkeithing with seventeen years in the RE – told Rees that the filling had been steamed out to well below the fuze chamber. He thought that it should now be safe to haul the bomb out of its hole and to complete the work in a back street while the main road was re-opened, allowing traffic to flow again.

Rees, a civil engineer by trade, had his doubts. He believed that the presence of the (50) might indicate that there was a second fuze, probably a (17), in that part of the bomb which remained buried. Munford, however, had anticipated that and had listened with a stethoscope without hearing any tell-tale ticking. Rees gave instructions for a clamp to be fixed to the bomb and a rope attached from this to a lorry.

'When everyone had cleared,' said Rees later, 'I issued an order to the lorry driver to take the strain on the rope, and while this was being done I heard a shout and commanded the lorry driver to stand fast. It was Serjeant Finney who shouted, and until then I did not notice that he was standing near me. He shouted because the rope was slipping on the lorry, and when the lorry came to a standstill he stepped forward to make the rope fast.'

Rees told the driver to take up the strain again, shouting to him: 'You are pulling!' 'All my attention was riveted on the bomb hole at this time, and as soon as I said "pulling" a huge flash appeared in the hole, accompanied by a thunderous explosion.

'I felt a tremendous blast to my face and was spun round. I realised debris was falling and managed to get under the front of the lorry. When I came out, I saw a body on the ground on the remote side of the lorry from where I had been standing and on further inspection I found five other bodies nearby.'

The other sub-section, which was near by, heard the explosion and ran to help. There was little that they could do. Some of them would have been working on the bomb but for the fact that they had had no sleep for several nights in a row and had swapped with others in the squad. By a fluke, in going to the aid of their friends they cheated death a second time when the UXB on the sea-front detonated in their absence.

Harry Vallance, a stonemason from Lockerbie, was one of those men, and the first thing he saw was Rees staggering about in shock, bleeding from the ears. Finney had also been wounded. Although both were to spend a long time in hospital, by a miracle they had survived. Munford, Henderson and five other men – almost all of them experienced NCOs – had lost their lives.

It was never established whether it was the (50) or a hidden time fuze that had detonated. Either could have been jolted into life by the pull of the rope, but what was less easy to explain was the presence of so many sappers so close to the bomb. The only explanation that Rees could give was that they had come out of cover to help when the rope had begun to slip, and he had not seen them as they were on the other side of the lorry.

Lacey went back to the billet, and kept everyone busy packing up the trucks in case there was another raid. When the siren sounded and the parachute flares started to drop for the third evening in a row he drove the section out to the Gower Peninsula. Near Blackpwll, he decided to turn off the road on to a golf course that looked safe enough, only for a soldier to appear out of the bushes and say that he was about to start lighting decoy fires there to draw the bombers away from Swansea. Lacey sought another refuge for the night.

Meanwhile, the town continued to burn. By the end of its blitz, 230 people lay dead, and several hundred more had been wounded. The following day, 22 February, all the sections in 8 Company, including Archer's at Cardiff, were brought in to Swansea to tackle the sixty-seven remaining UXBs, as was an RAF mobile unit and a squad from Birmingham. Accordingly, the Welsh capital was unprotected when later in the week it was raided heavily. Harry Vallance had had the grim task of trying to identify his former comrades from what remained of them. So many had died in Swansea that the mortuaries were full and buildings such as school halls had to be used in their stead. Vallance could never shake from his mind the images of the many dead civilians: 'I think it was one of the worst sights I ever saw in bomb disposal. The bodies were stacked the way you would stack wood – like a pyramid . . . They were a dreadful sight because they mostly turned black and they were bleeding.'

'The coffins of these chaps,' he remembered of those killed at Castle Street, 'were sealed and not to be opened. They were only filled with sand, in my opinion. The smallest chap had the heaviest coffin; the largest chap, the staff serjeant, had the lightest. I think that upset some . . . some chaps did give in, you know. They were transferred after these sights.' The police and undertakers informed the relatives of the dead men, but under the circumstances formal identification was impossible.

'All that was left for us to do,' wrote Lacey, 'was to collect the coffins and on the following day to take them to the railway station.' It was almost with relief that the Company was able to turn its attention a few days later to the havoc in Cardiff, which by early March was being choked by more than 150 UXBs. By the end of the month almost all of these had been neutralised, and 8 BD was moved from the front line to the calm of Oxfordshire for a much-needed period of rest.

14

THE HAND OF WAR

John Hudson had taken up his new post at BD Headquarters in Westminster only a few days before the disaster at Swansea. In the weeks that followed he compiled and analysed accounts of all the fatal incidents involving UXBs. Until then there had been no systematic attempt to understand why things went wrong, or to use that knowledge to make operating conditions safer.

Hudson's findings would underpin what would become his increasingly important work in bomb disposal. They would also dispel some of the myths that had grown up around BD, reflecting a decisive shift from its being a heroic struggle against the odds to a professional military organisation.

Fatalities could not be avoided simply by instructing sappers in the properties of particular fuzes. It was common knowledge that individual soldiers and bombs alike had their foibles, and what Hudson hoped to learn was which of these bore more responsibility for BD's casualties. His conclusions, and the other insights to be gained from similar data, are thought-provoking and rather surprising.

By the time Hudson finished making his analysis, in early June 1941, 150 BD personnel had been killed in the line of duty in

Britain. Another seventy-one had been injured. Some 60 per cent of the fatalities had occurred in the first four months of the campaign, as might be expected since that was when the bombing was heaviest.

Of forty-nine incidents studied by Hudson, the highest number – thirteen – had been caused by attempts to extract a (17). These had resulted in, on average, two deaths, usually an officer and an assisting NCO. Far more deaths, however, had arisen from accidents in which a (17) had not been left for four days or where a clock-stopper had not been used.

Hudson's evaluation of the figures led him to four deductions that contradicted the prevailing perception by section officers of the relative risks of their work. 'Ninety-three per cent of all accidents in Bomb Disposal,' he concluded, 'could have been avoided. Fifty per cent of the officers killed on Bomb Disposal have been killed while trying to remove (17) fuzes. NO officer has yet been killed (but thirteen sappers) on Bomb Disposal in an unavoidable accident (spontaneous explosion of Cat A).'

Furthermore, he calculated, 'in six months' work' – after which one could ask for a transfer – 'the chance of anyone being blown up is five hundred to one AGAINST, if safety precautions are taken as laid down in standing orders'.

Hudson was not saying that the job was safe. His point was that what did kill soldiers, by and large, was not the unpredictability of fuzes, about which little could be done, but carelessness, hastiness, complacency and fatigue, all factors that were often within sappers' own power to mitigate or to control. Of course, there would always be some cases of bad luck, and losses had at times been high, but there was no reason why they could not be significantly reduced in the future.

Whether this positive spin on their work did lift the spirits of those in the field is hard to know. What is certain is that it informed Hudson's own efforts, leading to an increased emphasis, in the training manuals and field handbooks that he wrote, on adhering to established guidelines. Safety distances were to be observed, lessons were to be learned, limits were to be set on the numbers employed on each phase of disposal. Combined with advances in equipment and with the fall in the number of UXBs now landing on British soil, from mid–1941 this more structured approach helped significantly to lower the number of casualties suffered by BD.

The data reveals a number of other statistics that contradict some of the general assumptions about who was most exposed to risk. Fourteen officers were killed between September and December 1940, and just six in the following six months. The respective figures for NCOs were eighteen and twenty-one, and for Other Ranks fifty-five and thirty-five. There had been about 1500 men on BD duty in September, with about 6500 by early January 1941. In that time some twelve thousand UXBs had been dealt with.

What is surprising about those numbers is that almost twice as many NCOs – representing a third of all casualties from January to June 1941 – had lost their lives as had officers, who were in theory most vulnerable to the dangers of fuzes. The reason for this may be either that the NCOs were increasingly given responsibility for supervising disposal parties, or that they were taking it upon themselves, now that they had some experience, to withdraw fuzes.

Taken in the round, the figures suggest two diverging casualty patterns. That for officers had been initially relatively high, but as much of the technology available was dedicated to making extraction safer their task rapidly became less hazardous and their losses fell. New equipment, however, could do little to alleviate the risks associated with digging down to a bomb that might be ticking. From early 1941, it was NCOs and sappers who were in fact in the greatest danger.

It is similarly interesting to note that, according to Hudson's research, the much-feared Zus claimed relatively few victims. Moxey and one or two others may have fallen prey to it, but once Stuart Archer had retrieved one at Llandarcy its threat was known.

Arguably, the bulk of the casualties that the Zus did cause stemmed – as at Swansea – from its preventing (17)s from being removed, meaning that they remained live for longer.

The other key determinant of risk was location. London, of course, had by far the largest number of UXBs, but it also had the most BD sections. Proportionately, at any rate, it was not the most dangerous place to be stationed, suffering about one death or injury per fifty bombs by the end of 1942. No. 16 Company, in Cardiff, lost only six men compared to the sixty-five killed in London, but it dealt with just 229 UXBs instead of five thousand. In Manchester, No. 10 Company had a similarly heavy ratio of casualties, about one per thirty-eight bombs.

In contrast, No. 2 BD Group, based at Tunbridge Wells, had by 1942 tackled 3434 UXBs with the loss of just one officer – Frank Martin – and sixteen sappers harmed, making service there four times safer than in London. The widest disparity of all was between Cardiff and No. 11 Company, that which covered Scotland, which had countered twice as many UXBs as No. 16 for six times fewer injuries.

There are probably several explanations for these differences. A lack of experience may be one: those areas that had relatively few bombs tended to have comparatively high casualties. One major disaster could also distort the figures. So the luck of the draw – where you were posted in BD, and what you were confronted with – had a part to play in your fate. But you made your own luck too.

Although the number of raids on London was to decrease sharply from the New Year onwards, paradoxically the proportion of UXBs dropped was to rise. Whereas they had constituted about 5 per cent of bombs used at the start of the Blitz, by the spring of 1941 that percentage had doubled. Over months in storage, the rain, snow and damp of a northern European winter had penetrated many fuzes.

The weather had not improved the Luftwaffe's temper either.

Like an ageing boxer determined to show he was still a con-
tender, when it did strike the capital it put all it had behind the
blows. There were two heavy attacks in mid-January. That of the
11th killed 111 people sheltering in Bank Underground station
when a bomb tore through to the escalator machinery room
before exploding. Some of the victims were killed by being blown
into the path of an arriving Tube train. The crater created outside
the Mansion House by the damage to the subterranean ticket
hall was so vast that the Royal Engineers had to be brought in to
bridge it, and it was not repaired until May.

The next major assault came in early March, but it was the
three that followed which were to scar London for longest. Those
of 16 and 19 April – the 'Wednesday' and the 'Saturday', as they
became known – were the two most brutal of the Blitz so far. In
the first, which lasted for more than eight hours, devastation was
unusually widespread and two thousand fires were started.
Eighteen hospitals, thirteen churches and sixty public buildings,
including St Paul's, were damaged or destroyed. Almost 1200
people lost their lives, a figure matched when the bombers
returned three nights later, and with even greater ferocity. In the
space of half a week some 148,000 homes had been wrecked or
battered.

Following these raids, Len Jeacocke was ordered to work on a
UXB that landed just a few yards from the base of the Victoria
Tower at the Houses of Parliament. It had been rated Category A,
and the section laboured hard to break up the reinforced roadway.
'We dug all day,' he remembered, 'but by evening we were
exhausted and had still got down only three or four feet. I
thought, Blow this, I don't know about the rest, but I'm going
back to Chelmsford for a pint, and come back on the first train
tomorrow.' He told the section to knock off, but to be back on
site by half past eight the next morning.

Having had a word with the station master at Chelmsford,
Jeacocke caught the fish train at dawn as it slowed down enough
for him to hop on. By eight o'clock he was back in Central

London. 'I was walking along the Strand towards Whitehall when there was the most almighty explosion. When I got to the site, the Houses of Parliament were still standing, but there was a bloody great crater in the road, the gas mains were on fire, the water mains were spouting. And all my tools were gone.' Had it happened half an hour later – or had the section worked non-stop as ordered to – then they would all have been killed. 'An officer did ask where we were at the time; I said that we'd gone for breakfast.'

Yet April's scale of destruction was exceeded just three weeks later, on the night of 10 May. Until almost the middle of the twentieth century, large swathes of London had remained remarkably Dickensian in feel, shabbier and more homely than the city known to those born after the war. It was a metropolis whose centre was still largely built of sooty brick and tile, not stone and concrete. Few buildings soared higher than the spires of Wren's churches, hand-painted signs sprawled across the upper façades of shops, and awnings sheltered passers-by from the rain. The servant classes lived in slum conditions around the back of the Mayfair mansions whose laundry they took in.

It was not simply the East End that was synonymous with poverty and crime. The building of Kingsway had begun to sweep away the fetid alleys around Covent Garden familiar to Thackeray and Wells, but still within hailing distance of the Royal Courts of Justice, the Bank of England and even the Houses of Parliament were mean rows of grimy houses where dwelled the poor and desperate. It was this Victorian way of life, as much as the city's most familiar landmarks, that was blasted by what was to prove the last great shock to the Empire's heart.

The FA Cup Final had been played that afternoon, and as Ballard Berkeley patrolled the West End in the late evening on his special constable's beat he could hear the newspaper vendor outside the Lyons Corner House shouting: '*Star, News, Standard* – Cup Final result!' Berkeley, an actor later known for his role as the Major in *Fawlty Towers*, took cover as the incendiaries started to fall.

'And then a prostitute came up from Piccadilly, up from the Empire cinema,' he recalled. 'It was a warm, clear evening and as the incendiary bombs came down she put her umbrella up and she was singing "I'm singing in the rain". The only rain was the incendiaries. It was quite extraordinary. I remember thinking at the time, I wish Hitler and Goering could have a look at this.'

Not everyone was so calm. It appeared as if the north side of Old Compton Street would be burned down, so Berkeley began to help the Fire Service to evacuate shops and a bedding factory that was alight. A six foot four Canadian followed him drunkenly along the road, and when Berkeley realised that the man was stuffing looted cheese, wine, whisky and cigarettes into his great-coat he moved to intervene. 'Then, suddenly, it happened. I had heard bombs, been near bombs. But when a bomb hits so close, you don't hear it. Everything stopped and there was complete and utter silences. It was so quiet – it was unbelievable.'

Berekely thought later that the explosion might have tem-porarily deafened him. A grey mist filled the street, which looked to him like a scene in the theatre with a gauze stage cloth in front of it. Then everything came to life – everything except the people who were dead.

'This was a four-thousand-kilogram bomb. It had dropped on St Anne's Church at the end of the street. I couldn't move. I thought that my legs had gone, but in fact I was up to my knees in rubble. As I was clearing the rubble I was thinking, Am I all right? I thought that I was, but then the most extraordinary thing happened. A tremendous panic came on me. I was suddenly scared out of my life, and I had to get away. I had to run, and I did.

'After a few yards, I was suddenly outside of my body and looking down at myself and talking to myself. I said: "Why are you running? Why are you frightened? This is your duty – you're supposed to do something here!" Then I seemed to re-enter my body and I became calm and cool and collected. I stopped run-ning, and went back.'

The blast seemed to Berkeley to have selected victims at random. Some people had survived untouched, while others had been annihilated. 'Then I saw a poor devil lying on the ground like a rag doll. It was my Canadian friend again. He couldn't move – he must have broken his back. I asked him if he was all right. "I'm all right," he said. Then he looked up and asked: "Can you give me a cigarette?"'

It was a cloudless night and there was a full moon. In the space of a few hours, some 550 aircraft dropped seven hundred tons of high explosive on the capital, starting 2200 fires. The spring air was filled with showers of sparks and the scent of charring timber. It was a horror film scored to the roar of collapsing buildings. A parachute mine landed on Bond Street, and an unexploded bomb blocked Regent Street. The Law Courts, Mansion House, the Public Record Office, the Queen's Hall and the Tower of London were hit, as were fourteen hospitals and five livery company halls. A quarter of a million books burned by moonlight in the British Museum. A third of the streets in Greater London were reported impassable, and 155,000 families left without water, gas or electricity. From Hammersmith to Romford, a pall of smoke shrouded the city.

From halfway up its spire, St Clement Danes started to burn, shooting out tongues of flame. So traumatised was the Rector by seeing its carbonised skeleton the next day that he died soon after of shock. Further along Fleet Street the Temple Church met the same fiery fate, as did the Wren-designed St Mary-le-Bow. The bells that had prompted Dick Whittington to turn around came crashing down. Westminster Abbey was less badly damaged, as was the neighbouring St Margaret's, the site of many a society wedding.

A total of 1436 civilians were killed that evening – the most by a single attack during the war – and almost 1800 men, women and children were injured. At eight o'clock the next morning Churchill's private secretary, Jock Colville, walked out from Downing Street to view the destruction.

'Whitehall was thronged with people,' he later wrote in his diary, 'mostly sightseers but some of them Civil Defence workers with blackened faces and haggard looks. One of them, a boy of eighteen or nineteen, pointed towards the Houses of Parliament and said, "Is that the sun?" But the great orange glow at which we were looking was the light of many fires south of the river . . .

'I stood on Westminster Bridge and thought ironically of Wordsworth and 1802. St Thomas's Hospital was ablaze, the livid colour of the sky extended from Lambeth to St Paul's, flames were visible all along the Embankment, there was smoke rising thickly as far as the eye could see. After no previous raid has London looked so wounded the next day.'

At the doors of Westminster Abbey were fire engines. '"There will be no services in the Abbey today, Sir,"' a policeman told him, 'as if it were closed for spring cleaning.' Flames were still spewing from the roof of Westminster Hall, the ancient heart of the Palace of Westminster, while the Chamber of the House of Commons had already been consumed.

'The bomb had fallen almost directly above the Speaker's Chair,' noted one shocked visitor that day, the MP and journalist Vernon Bartlett. 'A cloud of dust still hung over the place. The stone of the doorway into the Chamber . . . had been flaked and eroded in one night so that it looked as old and as weather-worn as the ruins of Ancient Rome.

'As I clambered up the hill of rubble I was suddenly confronted by a figure clambering up from the other side. There stood Winston Churchill, his face covered with dust, through which the tears that ran down his cheeks had made two miniature riverbeds.'

That same day, Len Jeacocke went with Captain Oswald Robson to Victoria Station, which was on fire and strewn with five UXBs. The highest priority was a bomb on the track. Some wagons were pushed over it, and it was these that took the force of the blast when it detonated two days later. The men removed another UXB that was ten feet from the entrance to the Underground station, and then dug out one that had landed on Platform 2. When the job was finished Jeacocke was granted permission for some brief leave.

'I got myself a ticket to Bournemouth to see a girlfriend. When I got out, the ticket bloke said there was an excess fare to pay, 1/9d, and I says I've just spent three days digging bombs off your Victoria Station. It didn't make any difference – I had to pay the one and nine.'

'In the main,' Jeacocke thought, 'we lived dangerously and we lived very well together. I've never found camaraderie like it. The spirit was great.' He worked with both Suffolk and Blaney, helping the latter to make safe a parachute mine in the accumulator shed at Stratford railway depot.

The early safety horns used to deal with parachute mines had been replaced by a gag that consisted of two short brass rods, one of which ended in a half-inch disc. This was inserted into the fuze, leading to a heart-stopping moment when the clock would begin to run before the lever of the escapement jammed itself against the brass plunger.

'That was the first operation,' recalled Herbert Hunt, who performed it on several mines. 'The right hand and fingers had to be held steadily pressing on the fuze. To remove pressure at this stage would cause the mine to explode. Then with the left hand, using the forefinger and the thumb, the securing pin would be inserted through the copper disc and into the fuze head and screwed up tightly. This would lock the gag and prevent the lever going over to detonate the fuze.

'On these operations, many things flashed before your mind and although the operation took a short time, it seemed like hours. In fact, after the fuze had been secured, feelings were so, that it was a great effort to let go of the gag, in case something might happen, the same experience as with delayed action bombs.'

Even once defuzed, the awesome potential of the mines could still induce fear. Hunt could vividly remember one occasion on which he had had to travel in a lorry loaded with two that he had helped a naval party to gag. As the vehicle began to move the mines began to emit an ominous clicking noise. Hunt spent the entire journey in a state of near terror, and not until much later did he learn that the sound had been made by the wholly harmless rotation of the mines' gyroscopes.

Parachute mines were used extensively in the April and May raids on London. One demolished Professor Andrade's research laboratory as well as personal effects, including his notes and apparatus. When another landed on a terraced street in Pimlico, twenty-seven out of thirty houses were flattened, and one man was blown a quarter of a mile by the explosion. A third drifted intact on to the railway lines of Hungerford Bridge, leading into Charing Cross.

Dozens of incendiaries dropped on to the station had started a large fire in the timbers of the bridge. The flames got to within ten feet of the bomb on the tracks before they were subdued. When Sub-Lieutenant Ernest Gidden of the RNVR arrived shortly after dawn on 18 April he found that the heat that had been generated had fused the mine to the live rail. Breaking off a piece of glass from a carriage window, Gidden used this as a mirror to check the exact location of the timer on the reverse of the bomb. Unfortunately, it was so close to the rail that he would need to free the mine and turn it over in order to insert the gag.

The UXPM on the bridge had stopped both the overground and underground trains from running and many buildings, including the War Office, had been evacuated, so Gidden was under pressure to clear the danger as soon as possible. Nonetheless, so distorted was the casing that it took him six nerve-wracking hours of levering and bashing with a hammer and chisel before he could gag the fuze.

He was helped to remove the mine by one of the stalwarts of 5 BD Company, Lieutenant Nevil Newitt, who had commanded the detachment that had steamed out the Regent Street UXB at the start of the Blitz. Newitt would later become one of Hudson's colleagues at Romney House, BD's headquarters, although no one quite knew how he had ended up in bomb disposal: not, that is, unless being an undertaker was thought to be a suitable qualification. Gidden was awarded the GC for his courage on the bridge, adding to the GM that he had won three months earlier.

London was not the only victim of the springtime raids. Belfast, Hull, Plymouth, Bristol, Southampton and Clydebank were also mauled. From time to time, even small ports such as Falmouth, Bournemouth and Fraserburgh were hit. In Ramsgate seventy-one people were killed and 8500 houses damaged by raids, while after London nowhere was bombed more frequently during the war – on ninety-seven days and nights – than Great Yarmouth.

Merseyside suffered more grievously than any other place outside the capital, being as it was the principal port of entry for supplies from America. Parachute mines were deployed here, and in an attack on 28 November 1940 one of these fell through the crown of a gasometer at the Garston gasworks. It landed inside without exploding, but all around it were some two million cubic feet of gas.

Work was paralysed in the upper reaches of the Mersey. Six thousand people were evacuated, three docks and their yards were closed, as was the railway, while the entire gas supply to east and south Liverpool was threatened.

Harry Newgass was a funny sort of Englishman, a well-to-do Jew raised as a landowner and fond of country sports. His small build and eyeglass did not suggest a man of action, but he had fought with the Royal Artillery in the First World War, and although nearly forty-five had swiftly joined the RNVR for the Second.

On arriving at the gasometer he saw that the mine was standing part immersed in about seven feet of oily water. One of the fuzes lay against a brick pier supporting the roof, so the whole bomb would need to be delicately revolved before work could begin. Newgass insisted on undertaking this alone and entered the holder through a hole cut in the roof. Should anything go wrong, he had absolutely no chance of escape.

To prevent himself from being poisoned by the atmosphere, Newgass wore primitive and cumbersome breathing apparatus. The water had been pumped out, and he was able to lash the ring at the base of the parachute to the pillar against which the mine rested, thus making it less likely to topple over. He then heaved sandbags together to make a small mound on which he was able to pivot the heavy mine using a long crowbar. This would have been a hard job even for a bigger man, and Newgass was handicapped further by having to climb back up to the top of the gasometer to change his oxygen cylinders every half hour.

Only after six such trips and three hours' work was he able finally to get at the fuze, which needed to be removed by hand and proved to be stiff. Once this was out he was able to extract the magnetic primer and then the clock itself. Working in a sappingly hot and foul environment, by the finish Newgass was, as stated in the citation for his GC, 'completely worn out'.

Les Clarke was stationed with 3 BD Company in Liverpool at the height of the attacks on the city in May 1941. Although born in Edinburgh, he had grown up in Chatham and at twenty-two was rather dismayed by the outbreak of war. Not only would it interrupt his promising career in civil engineering, but he had also just invested in a house for himself and his new bride.

By early 1941, Clarke had joined an RE Field Company, but the rumour was going round that they were bound for India. 'I didn't fancy dying in India,' he recalled, 'and I didn't think I'd like the food. So I thought: let's volunteer for bomb disposal. I spoke to my friend Corporal Mumbery, who came from Tonbridge, and he said: "We'll get killed." I said: "We might as well get killed in England."'

Among the early jobs that he worked in Liverpool was a Category A UXB at the Bryant and May match factory in Garston, as well as several on Scotland Road near the northern docks. He was with a steam sterilising section, but such was the urgency to clear bombs that they began to take out fuzes rather than wait for the laborious emulsifying process to run its course.

He was lucky, but inevitably others were not: 'We lost lads, but there wasn't a lot said. It was rather tragic really. They were with you one minute, and gone the next.' News of their deaths was kept from the public to avoid affecting morale.

'If somebody happened to get blown up, and he was sleeping in the bed near you, you knew he was gone because his gear was gone. And somebody new would come.' During the eight nights of sustained bombing at the start of May that would claim 1700 lives on Merseyside, the house in which they were quartered was bombed. 'There were fires raging,' he remembered. 'A lot of women – and men – were shaken . . . I suppose there was hatred towards the Germans.'

The section was forced to move out to Huyton – 'the trams used to stop for us, they'd see our [shoulder] flash and we'd never pay, took us into Liverpool' – from where they did their best to keep up spirits. For several days they searched for a bomb reported as having fallen near Aintree, but without success. 'So we got a bomb we'd taken out previously, and we took it at night-time and stuck it down the hole.' The next morning, they pretended to resume their search, watched from a distance by an appreciative audience.

'The crowds were all roped off,' he grinned. 'When eventually the crowd got big, we produced this bomb, and the crowd cheered. I had a chap in my squad who was a good cadger. He went round and made a collection. We went to the first pub we could find and spent it all.'

Although no one knew it, the Blitz was nearly over. A few hours before the last great onslaught on London, the Earl of Suffolk drove out to one of the main bomb cemeteries, Erith Marshes. Local residents had for some time been campaigning to have the work shifted elsewhere because of the noise, and BD had now begun to tidy up its handiwork. The few remaining bombs were to be moved to Richmond Park, but Suffolk had been told that there was still a 250-kilogram at the site with (17) and (50) fuzes intact. Examples of these were still in short supply, and they could usefully be recovered for testing. The bomb had fallen in the autumn and had become such a fixture of the place that it was known to the sappers as Old Faithful.

Since the turn of the year, Suffolk had continued to lead the MOS's experimental squad. In folders that he had brought from France he recorded in precise detail his tests with the new plastic explosive, RDX, on what he called 'Satan's babies', or medium-sized bombs. In February, however, he was told to stop these after the Navy and Air Force decided that the method was not safe for use in the field.

BD's commanders continued to express their faith in his work, but many of the sections that encountered Suffolk had doubts about the way in which he went about it. One officer accused him of carelessness after a five-hundred-kilogram bomb half wrecked a house while Suffolk was trying to neutralise its fuze with gelignite. The Earl did his best to defend himself against the charge that his team were meddlers. A few weeks later, however, a detachment of the East Surreys stationed in Richmond Park complained about several violent explosions set off by Suffolk that had, without warning, sprayed red-hot metal close to the guard room and officers' mess.

William Wells would sometimes see Suffolk, dressed in a Stetson and airman's boots, around Company Headquarters. He got closer to him at a demonstration of the thermite burning-out technique. On this occasion the Earl was wearing a white kid flying helmet and fur gloves, and was perched on a shooting stick. For some months Wells's squad had been using a home-made method of burning defuzed bombs with a paraffin rag, and he and other officers gave Suffolk's scheme a cool reception: 'No one seemed very impressed by the new method. In fact, the view was advanced that some of these scientific wallahs were so scientific that when they wanted to open a bottle of beer they instinctively applied themselves to cutting a hole in the side instead of removing the stopper.'

To soldiers with a hard-won awareness of the risks of bomb disposal, Suffolk's exuberant style was at odds with the cautious
approach that they had learned to adopt. John Hudson was
another who had misgivings: 'He was a very colourful, very odd
man. He used to have a pistol in a holster under his arm and when
he wanted his van he would fire his pistol twice in the air, that
sort of thing. He had a very pretty girl as a sort of secretary, we all
envied him her . . . We didn't care for Suffolk at all. He wasn't disciplined, and he wasn't one of us.'

Yet those sappers who worked more closely with Wild Jack had
a different view. A section under Lieutenant Richard Godsmark –
who had led Harry Beckingham's trio at Sheffield in the first
days of the war – had been put at Suffolk's disposal, and while
doubtless they saw it as a cushier job than most in BD, they
'formed the friendliest relations'. He took them to dine at
Kempinski's in the West End, still in their muddy jerkins, and
pleaded to retain 'my little squad' when at one point it seemed as
if he would lose their services. Writing to Taylor, as a former
Guards officer Suffolk praised Godsmark's leadership and noted
that 'while he shows no signs of timidity he observes a very proper
degree of caution for himself and his men'. For their part, the
men clubbed together to present Suffolk with a silver cigarette
case inscribed with their names.

On Monday 12 May, with London still reeling from the bat-
tering it had taken two days earlier, the section made its way out
to Erith. Suffolk had telephoned at about lunchtime to say that
the (17) was ticking, and he had sent for a Mk. II KIM clock-
stopper. By a quarter to three this was in place, as was a
stethoscope, and preparations began to be made to sterilise the
bomb. Two sappers were sent to a nearby ditch to collect water
for the boiler. And then there was a bright white flash, as if a star
had burst.

There were fourteen people around the bomb when it
exploded. No trace at all could afterwards be found of four of
them. Three vehicles were set ablaze and the doors and name

boards of Suffolk's research van were wrenched off and flung sixty feet by the blast wave. Its engine block was perforated by fragments of steel.

Among the dead were Staff Serjeant Jim Atkins, who had been operating the stethoscope, as well as Sappers Jack Hardy and Reg Dutson, and Driver David Sharratt. Another sappper, Bert Gillett, died in hospital the next day. Six further soldiers had serious wounds or were in deep shock.

Beryl Morden died in the ambulance that had been called to the scene. Of Fred Hards nothing remained. Some fragments of clothing and feet were identified as belonging to Suffolk, but in essence all that survived of him was the silver cigarette case.

Earlier that afternoon, Herbert Gough had ventured out to the eastern edge of the capital to see Suffolk at work on the bomb for himself. 'I was always a bit nervous about him,' he said later. 'He gave the appearance of being slapdash at times, although he had my confidence: but I was uneasy about my friend, so motored down to see how he was getting on.'

When Gough arrived at the desolate marshland he reproved himself for having doubted Suffolk's professionalism. 'He was going through all the motions of safety, taking every precaution he should. When I reached my office I was told that it had gone up, and Suffolk and all his team with it.'

Having sent a warning telegram, and then having frantically tried to confirm with rural post offices and police stations that it had arrived, early the next morning Gough drove down to Charlton to break the news to Suffolk's wife. 'A painful interview' ensued. The Countess had known little of her husband's work. She had seen him only the day before, when he had managed to spend the weekend with her and their three sons, the youngest still a baby. Now she would be left to bring them up alone, and in somewhat straitened circumstances. Charlton had been requisitioned for use as a military hospital and the family was living in just three rooms. The War Office had yet to pay any rent and Suffolk's mother had recently stopped his allowance on the grounds that he now had a job with the Civil Service. During the next few months, Gough was to try hard to get Lady Suffolk a war widow's pension, but his plea was met with an unsympathetic response from officials who found it hard to believe that anyone living in a stately home could be in real 'pecuniary need'.

Although the nature of Suffolk's work, and the manner of his death, remained secret in wartime, some public tributes were nonetheless paid to him. Gough wrote an obituary for *The Times* that mentioned only 'special duties', although in Parliament Lord Clanwilliam revealed that he had been 'destroying unexploded bombs'. Taylor saluted him as 'a very brave and gallant gentleman'. Perhaps more touching because of their lack of formality were the letters of condolence that came to Gough from technicians who had collaborated with Suffolk on his experiments. Among them was the chief engineer of the London Power Company: 'My workshop staff all feel they have lost a pal', he wrote, recalling Suffolk's drive, charm and 'disregard of danger', as well as his assistants, 'the little Cockney and the fluffy-haired lady'.

Gough took their deaths hard. He felt in some measure responsible, and began a campaign within Whitehall for 'exceptionally favourable and compassionate treatment' for the Trinity's dependents. Neither Morden nor Hards's families were well off, and though she was unmarried he had left a widow and young children. Gough also made strenuous efforts to have all three awarded a posthumous honour.

Yet not all shared his view that the disaster at Erith was an unavoidable tragedy. Recriminations and rumours had started to fester within BD. Many were angry at the number of sappers who had been killed. Why, it was asked, had they been so close to a ticking bomb, in contravention of all the regulations? The finger was pointed at Suffolk's maverick approach.

The news of his death came as no surprise to William Wells. He heard that Suffolk had been using acid to try to destroy the clock of the (17), and that Atkins had been only five yards away with the headphones. 'Some officers,' he observed, 'particularly those who had been on operational duty since the early days of bomb disposal, were very critical of this heavy loss of life.'

'It seemed a blunder on the part of the higher authorities to employ Lord Suffolk on bomb disposal experiments. He was a courageous man, but too adventurous, too much the showman for the cold-blooded business of bomb disposal experimental work. This was wholly the province of the backroom boys of the NPL, the RAE and some industrial laboratories. These boys were never in the limelight . . .'

John Hudson was similarly disparaging of Suffolk: 'We were all very thankful to have him off our chests, but we were very sorry that he took a lot of our chaps with him . . . He was an amateur, the one and only amateur bomb disposer . . . I mean, to have people standing around watching you in bomb disposal just wasn't done.'

What exactly Suffolk was doing at Erith just before the explosion was never established. Hudson believed that he might have been trying to show the sappers how to tell if there was a Zus under a (17), and that this had gone off as he was withdrawing it. But, like all the criticisms of Suffolk and suspicions that he was treated favourably because of his title, this was to mistake the externals of his appearance and manner for a lack of rigour in how he conducted his work. It also overlooked the fact that the sections and the experimental squad were governed by different imperatives. The objective of the former was to render UXBs safe without exposing themselves to unnecessary danger. The work of the latter, as Gough had told Taylor, often consisted of tasks undertaken precisely because they were considered too dangerous for the Army – such as recovering fuzes – and in such circumstances risks had to be run.

It is known that Suffolk was endeavouring to recover the (17), but that the sterilising process had begun indicates that he had not yet attempted to extract it. His previous record of caring for the safety of Godsmark's men also suggests that he would not have been reckless with their lives on this occasion. Everything points to the fact that the explosion was entirely unexpected. Since the UXB had been dug up seven months earlier, any charge in its (50) would have long since leaked away, so the cause must have been the (17). Why it had started to tick at all is impossible to know, but in any case Suffolk had quite properly tried to quiet it with the clock-stopper. He must have thought that this had proved effective, and so the soldiers had gathered around to set up the boiler.

Whether the batteries on the stopper were not fully charged, or whether the noise of the lorries masked the ticking in Atkins's headphones, can only be speculative explanations. The truth is that such things happen, even to experienced officers such as Blaney and Ash, and that weaponry often shows perverse ingenuity in waiting until it can cause the maximum sorrow and pain.

In mid-July Lord Beaverbrook, then Minister of Supply, wrote to the three families. Jack Howard had been awarded the George Cross, but despite Gough's lobbying his two companions in danger were to receive only commendations for bravery. The official line was that the George Medal, for which they had been nominated by Gough, was not awarded posthumously. In the memories of those who knew them, however, as in death, the trio were inseparable. 'Up went the Earl of Suffolk in his Holy Trinity,' wrote Churchill, inspired by their exploits to think of them as latter-day Pilgrims, 'but we may be sure that, as for Mr Valiant-for-Truth, "all the trumpets sounded for them on the other side".'

15

KILLCAT CORNER

On 30 September 1941 an exhilarated Stuart Archer sat down at his desk to pen a note to his wife Kit, who was pregnant with their second child. Then he gave it to a dispatch rider to take round to their billet. 'Look at today's *Times*,' it read, 'and you'll find that I've won the George Cross.'

More than a year after his heroics at the refinery (and a year to the day since Bob Davies had been given the first GC) recognition of Archer's work in South Wales had finally come. Yet he was not to know that while the process had been a lengthy one it had also been much vexed. In fact, his award – the first for BD since the spring – had reignited a long-running and tetchy debate in Whitehall about the decoration, which would only be settled by intervention from the Palace.

The quibbles had started almost as soon as the St Paul's awards were announced in October 1940. Churchill had been keen that many others for civil bravery should follow, and was aggrieved to find that by December 'how very few George Medals have been issued'. There had been sixty-five so far; 'I had hoped there would be ten times as many.'

The Prime Minister believed that if gallantry awards were

granted more plentifully morale would be raised. The Civil Service, in the form of the George Cross Selection Committee, thought otherwise. Even by the time the second batch of GCs and GMs was due to be agreed, in November, there were already signs of trouble.

'It was clear,' considered the Committee's chairman, Sir Horace Wilson, Permanent Secretary at the Treasury, 'that the grant of the Cross on the scale suggested would make the awards far less valuable than had been the intention . . . it had always been found necessary that there should be great scarcity in the highest awards.'

From his point of view, the problem was that the Services were putting forward too many bomb-disposal related recommendations for the GC. (Moreover, three-quarters of the medals approved so far had gone to servicemen, when the Royal Warrant establishing it decreed that it was intended 'primarily for civilians'.) For Wilson, the comparison to be made was with the Victoria Cross, only thirteen of which had so far been awarded during the war. Yet he was faced with twenty-three recommendations for the GC from the Services alone, not to mention those from the civil authorities, such as the regional civil defence commissioners.

This led to some rather undignified horse-trading. Pressure was put on the Air Force to reduce its nominations from five to just one – Moxey – and the Army brought its down from ten to three. The Navy, however, refused to budge from its list of eight, or to trim a clutch of names put forward for the GM, also largely for mine disposal.

The royal warrant had ordained that it should be awarded 'only for acts of the greatest heroism or the most conspicuous courage in circumstances of extreme danger'. Wilson's intent – that the benchmark should be set high if the GC were not to be cheapened – is understandable enough, but his judgement was awry. The reason that so many awards had been proposed was not mere pot-hunting by the Services but a reflection of the vast volume of bombs that had been dropped. Moreover, comparisons with the number of VCs were redundant when opportunities to fight the enemy had latterly been restricted by the retreat from the Continent. Indeed, Davies had been awarded the GC precisely because the location of the St Paul's bomb had ruled out the VC.

Davies's medal should have confirmed bomb disposal's status as being worthy of the highest available decoration. Yet not only did the Committee believe that his award had opened the floodgates too wide, but it also remained unconvinced that the risks run by BD were sufficient for such an honour.

In May 1941 it met again to review the requirements for the medal. Sir Alec Hardinge, the King's Private Secretary (and the recipient of a Military Cross), argued that 'in his view the majority of the recommendations . . . were not up to the standard of the Victoria Cross'. Meanwhile Admiral Lord Chatfield, who at Churchill's request had been instrumental in instituting and publicising the GC, felt that 'the large number of George Crosses given for bomb disposal, however right in its original conception that step may have been, is now inevitably leading to the GC getting into a different category from the VC'. Military men preferred their valour to be short, sharp and bloody rather than of the nerveless kind required by BD. Moreover, Chatfield continued, as 'the number of casualties are not believed to be very heavy', those engaged in it would be 'more appropriately rewarded in some other manner'.

General Taylor was asked to produce statistics about casualties per bomb in order that the Committee might gauge how dangerous the work was, but even these did not alter its opinion. Unfavourable comparisons were drawn with the risks run by submariners and by the airmen bombing Berlin, while Admiralty officials hinted that the War Office standard for the GC was 'less consistent and generally lower' than theirs.

Surely, argued one senior army officer, awards 'should be judged on the degree of gallantry, not upon any statistical calculation', but the civil servants were not to be swayed from their quotas. The fact was that its top echelons never fully appreciated what had been demanded of BD and how great its contribution had been to seeing off the Luftwaffe. Viewing as they did a decoration in terms of the boost it gave to public morale rather than as a reward to an individual for self-sacrifice, once the Blitz was past the Committee choked off the flow of GCs for bomb disposal.

In the first six weeks of that battle for Britain, the Royal Engineers had won a dozen George Crosses for BD. (Half of them went to men over forty, and three-quarters to those with professional qualifications; the work demanded experience and self-possession, not the reckless charge of the VC.) Further medals would soon go to the other pioneers in the fuze war: Len Harrison and John Dowland, the airmen who had been the first to brave a live bomb; Arthur Merriman, the lone retriever of fuzes in those early days when so little was known of them; Eric Moxey and Max Blaney, who had died trying to fathom their secrets.

By the time that Archer's record came up for consideration in July 1941, the Committee's mood had hardened. Taylor's recommendation focused on Archer's recovery of the Zus at Llandarcy the previous September, lauding his 'cold-blooded courage and dogged tenacity of purpose under appalling conditions'. Yet the Committee was now far less inclined to award the GC for bomb disposal work, even though Archer's exertions had been at the height of the Blitz. It rejected the nomination, deciding to give him the George Medal instead.

It was Hardinge who then had a change of heart. Writing in his capacity as the King's representative to the new committee chairman Sir Robert Knox – another keen student of casualty ratios – he noted that Archer had dealt with two hundred bombs and could not 'help thinking that [he] is deserving of the George Cross'.

Hardinge's intervention swung the others round and, belatedly, Stuart Archer received the recognition he merited. In November he travelled up from Thame, where he was now largely occupied with administrative tasks, to Buckingham Palace for the investiture. (Though he was brave, Stuart was conventional: he took his mother with him, rather than embarrass the King with the unseemly sight of his heavily pregnant wife.)

Just sixteen months earlier, he had driven down, side by side with Edward Talbot, to Cardiff and an unknown destiny. For both, it had been to win the GC, but only a few days earlier, Talbot's luck had finally run out.

Posted to Malta, he had tried to escape the strain of bomb disposal by accompanying an RAF crew on a routine flight over Sicily. Their Blenheim had collided with another from the same squadron, probably while trying to evade flak, and all aboard both aircraft had been killed. Archer read of Talbot's death in *The Times*. 'I happen to know that his bravery was not of the unimaginative animal variety,' wrote Talbot's commanding officer of his BD work in his obituary. 'He could be just as frightened as the rest of us, but it did not show.' The Selection Committee might have thought otherwise, but that was true courage.

In terms of casualties, it was Britain's civilian population that had borne the brunt of the war thus far. By the time Hitler launched his invasion of the Soviet Union in June 1941, marking the end of his attempt to conquer Britain, more than forty thousand had been killed by bombing, half of them in London. It would not be until 1942 – almost three years into the war – that the figure was surpassed by deaths of those in the Armed Forces.

At least another fifty thousand civilians had been so injured that they required hospital treatment. Some two million houses had suffered damage, and in Central London only one in ten buildings had escaped unscathed. Almost 190,000 bombs had fallen, twenty thousand of them UXBs.

Although the Luftwaffe, in an attempt to restore itself to favour, had portrayed the campaign as a strategic success, one that had weakened Britain enough to allow Germany to open a second front, it had patently been repulsed. 'He knows he will have to

break us in this island, or lose the war,' Churchill had said of
Hitler's aspirations in the autumn. They had not been realised.
Given the giant scale, the diversity and the widely scattered sites
of British industry, the Luftwaffe's task may always have been
impossible to accomplish, but its failure had had the effect of
rekindling a fighting spirit that many had doubted still existed.
Wars are attritional, and are not won without the capacity to
endure reverses. The Blitz had burnished the dull armour of the
British temper. 'Looking back,' thought even the sceptical John
Strachey, '. . . the people of London and the Civil Defence personnel had done quite well. They might have done better, but
they might have done a good deal worse.' Amid the ashes flowered
the wild pink of fireweed.

Although they ceased to be a nightly occurrence, the raids did
not stop completely. Coastal towns were hit by tip-and-run
attacks, as they were known, while Birmingham, Manchester and
London endured several heavy bouts of bombing in the summer
and autumn. The threat from these was reduced by improving
radar coverage, and by the introduction of nightfighters such as
the Beaufighter and later the Mosquito, equipped with airborne
radar that could detect German bombers. Scientists also learned to
counter the beams that guided the Luftwaffe by transmitting false
signals of their own which confused the German navigators and
their equipment, causing them to drop their bombs before they
reached their target. The successful institution of such electronic
warfare lightened BD's burden, but there was still work for it to
do, not least in anticipating any new types of fuze.

Three weeks after Suffolk's death, Herbert Gough decided that
there was still a need for an experimental squad, but even though
it would be working to the MOS's instructions it should be
manned wholly by the Army. Gough and his DSR had already
lost too many volunteers. The officer nominated by BD HQ to
run the new set-up was John Hudson.

'The work was all done in our own premises in Richmond
Park,' he recalled. BD's experimental station was at Killcat Corner,

Herbert Gough, chairman of the Unexploded Bomb Committee, director of scientific research at the Ministry of Supply and unsung hero of the fuze war. *(National Portrait Gallery)*

Herbert Rühlemann (centre), the German engineer who revolutionised aerial warfare by inventing the electric fuze, which was more reliable and versatile than its mechanical predecessors. *(Elga La Pine)*

Cornelius Stevens, RAF bomb disposer and inventor of the Stevens Stopper, which neutralised time fuzes by jamming their mechanism with a sugary solution. His son Barry (right) was killed in 1940. *(Courtesy of Paul Hughes)*

It fell to naval experts such as Harry Newgass (left) to tackle giant parachute mines dropped on cities. He was awarded the George Cross for tackling one in a Liverpool gasometer. *(IWM)*

Just before the war, Len Harrison reassembled for the RAF one of the earliest German fuzes brought to Britain. He later received the George Cross for defusing the first live UXB. *(IWM)*

Anti-personnel 'butterfly' bombs were sensitive to disturbance, but could safely be detonated at a distance by a tug on a rope looped round them. *(Topfoto)*

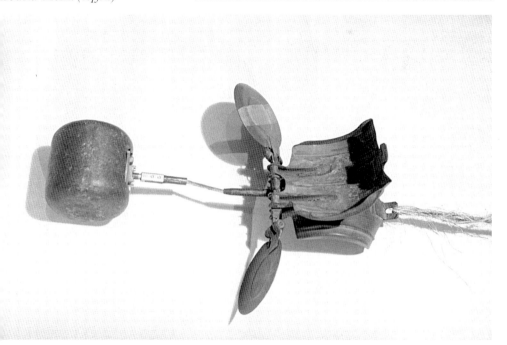

Winston Churchill surveys the devastation caused to the House of Commons on 10 May 1941 by the heaviest, and the last, attack of the Blitz. *(Getty)*

The King and Queen visit an area hit by a raid. The George Cross, which became ynonymous with bomb disposal, was named for the King, but the idea was Churchill's.

(Topfoto)

Scraping out by hand the explosive filling of a UXB was a method used to make safe a bomb when it was thought too dangerous to attempt to remove its fuze first.
(Crown Copyright)

Scientists devised ingenious machines to deal with UXBs. This Stelna trepanning head cut a hole in the steel casing of a bomb to allow steam to liquefy the explosive.
(Crown Copyright)

The spindly form of a 'Freddy' automatic fuze xtractor in position on a bomb. Its pistons were driven by bulbs of compressed air of the kind used to make soda vater. *(Courtesy of Steve Venus)*

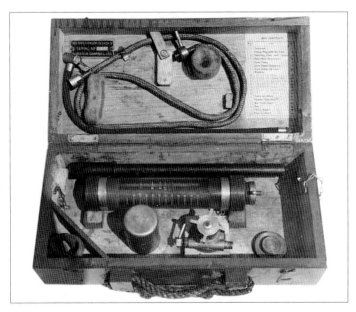

A liquid discharger kit, which utilised compressed air to force a cocktail of methylated spirit, benzene and salt into a fuze to conduct way its electrical charge. *(Courtesy of Steve Venus)*

Listening for the ticking of a time fuze in Normandy in 1944. Electric microphones and headphones were developed to replace the doctor's stethoscope used for this earlier in the war. *(Courtesy of Steve Venus)*

Doodlebug: a V-1 at its launch site across the Channel. It took bomb disposal experts more than a week to defuze the first rocket that landed intact. *(Getty)*

between the Roehampton and Robin Hood gates. 'We were given a series of cow sheds, which I ultimately blew up by mistake at the end of the war.'

Hudson's chief colleague and contact at the DSR was a twenty-six-year-old New Zealander, Bob Hurst. A physical chemist, he had been working on a doctorate on plutonium at Cambridge until recruited into the MOS. Hudson's role was to find out what the sections needed, and Hurst would have it manufactured. Hudson would then have to ensure that it worked.

'I was the chap who'd go and try it out in the field. It was always my custom . . . that if we had a new piece of equipment to introduce to our chaps then I would go out into the field and use it myself on a live bomb, not once but twice. Then I'd sign the instruction and if anyone got blown up afterwards it wasn't my fault because I'd done it twice and it worked. It meant our chaps came to trust us.'

Many of the scientific developments were demonstrated to a high-level audience in early June 1941. Among those watching in Richmond Park was Sir Alan Brooke, the Commander-in-Chief Home Forces, Harold Macmillan and his minister at the MOS, Sir Andrew Duncan, and Lindemann. After months of trying by Gough, it was the first time that any of them had been persuaded to come to see what BD actually did.

The great breakthrough at the DSR's laboratories in recent months had been the Liquid Fuze Discharger (LFD). This was an extension of Edward Andrade's tests in early 1940 that had aimed to introduce a liquid into the fuze pocket to dissolve the trinitro-phenol booster pellets. It had since been found that, at the same time, the solution penetrated certain kinds of fuzes, notably the motion-sensitive (50). This offered the possibility of being able to discharge its capacitors without having to risk moving it.

The challenge had been in determining what liquid might do this best. It was not until December 1940 that, following a chance conversation, the answer was supplied by a technician with whom the R&D Sub-Committee collaborated, John Davy of the

Glasgow optical instrument makers Barr & Stroud. The cocktail that would immunise the (50) proved to be one part pink methylated spirit to two parts benzene, 'with a little salt dissolved in it to give the necessary conductivity'. Benzene was used as it was a good solvent of any wax and grease that had accumulated around the charging pins, and would allow the mixture to penetrate the interior of the fuze.

As with the safety horn, an apparatus was rigged up to inject the liquid past the oiled washers around the fuze's plungers so as to saturate its interior. First, the fuze was pulled upwards a little to open up the pocket. Then a collet attached to a hand pump was fixed to the fuze boss, a gas-tight seal was made around it with petroleum jelly and a pint of the liquid was suspended above the bomb in a container, like a hospital drip. Once a controlled pressure of compressed air – indicated by a manometer – had been built up with the pump, a measurable volume of the liquid would be forced into the fuze when a valve was opened. The condensers could then be left alone to discharge themselves within about thirty minutes.

As a piece of kit, the LFD was lighter and more reliable than the earlier steam discharger, which required batteries, and by June 1941 was widely in service. Not only did it neutralise the menace of the (50), in turn allowing UXBs with (17)s to be moved much more easily from important sites, but it also worked against ordinary impact fuzes and thus superseded the Crabtree discharger as well.

Although Andrade and Merriman had developed the first LFD equipment, much of the work of refining it had been done by Wing Commander Cornelius Stevens, who had succeeded Lowe and Moxey as the key figure in RAF bomb disposal.

The son of a schoolmaster, the fourteen-year-old Con Stevens had left Birmingham in 1906, to enlist in the Navy. He had risen up through the ranks of the Royal Naval Air Service and later the RAF before joining T Arm 4, its BD division, in 1940 as one of its senior experts. By May 1941 the number of Air Force personnel engaged on UXB work had risen to eight hundred, and they had defuzed some 1600 bombs. Both of Stevens's sons were to follow him into the family business: James was to win the George Medal in 1944 for dealing with bombs on a crashed aircraft; his brother Barry was killed in 1940 when his aircraft went down in a snow-storm during a training flight.

The success of the liquid desensitiser had led Stevens to think that its principle could be adapted for use against the (17). The clock-stopper was proving effective, but it was a cumbersome object and its field was less potent against the thicker walls of larger bombs, a higher proportion of which were starting to be dropped by the Luftwaffe. What was needed was a method of ren-dering the fuze permanently safe.

To do that, it was not enough just to discharge its energy; the wheels of the clock had to be immobilised. It was by now known that this could be done with a viscous substance – Stevens favoured brine, made by dissolving three pounds of salt in a gallon of water – but the difficulty was to be sure from the outside that the cogs had been jammed. Stevens's solution, like most clever ideas, was simple. Instead of using a pump to inject a liquid into the fuze, he would use one to create a vacuum to suck the solu-tion in. In this way, all the air in the fuze would be replaced by fluid and the mechanism stuck fast.

Early tests on (17)s proved fruitless as they were airtight, but Stevens had faith in his method. Then, from early 1941, a new standard clockwork fuze was brought into service, the (17A). This proved less well-sealed than its predecessor, but although Stevens demonstrated that his technique was effective against it the UXBC refused to authorise the use in the field of the Stevens Stopper for another three years. Although the S-Set, as it was known, was also shown to work against the (50) it was feared that the similarity in method with the LFD might lead to confusion.

Only in 1944 when problems emerged with air locks in fuzes that could not be cleared by pressurisation from the LFD's pump, was the S-Set distributed to the Army and the clockwork fuze definitively mastered. As early as 1942, however, Hudson had successfully tested a version of it on a (17)/(50) combination UXB at Bristol.

One reason for that trial had been a change to the constituency of the LFD solution. It had been discovered that the viscosity of brine declined as its temperature rose, raising the possibility that the balance wheel of a fuze might suddenly free itself, for instance if warmed by steam during the sterilising process. Stevens therefore replaced the salt with sugar in both the LFD and the S-Set.

The various ingredients needed to be prepared on site, as Les Clarke remembered doing in Liverpool: 'To stop the fuze, apart from the magnet we used a solution of water and sugar mixed. We forced that by a bicycle pump into the fuze. The sugar ran low at times, and we found out that some of the lads were taking it for the tea. When the authorities heard about this, they sent us down a brown treacly fluid which was no good for tea. The lads tried that.'

Brine also interfered with another invention that Hudson had tested on the Bristol UXB. When the S-Set was used, it was seen that as well as filling the fuze it also flooded the Zus. Often this so slowed its striker that it failed to fire the (17) if this was then removed, yet it was not a wholly reliable state of affairs. Hudson realised that what was needed was a substance that would set solid once mixed with the water in the viscous solution introduced into the fuze. The answer was, as so often, found in a domestic setting: the pink powder used by dentists to make quick-setting putty for taking impressions of teeth.

Hudson had already had trouble with the (50) in the Bristol bomb even before he got to its (17). Using a new extractor developed by the tireless Ken Merrylees – a series of linked rotating screws worked from a distance by a long cord – he had tried to remove the (50) from eighty yards away, in a crater, only for it to catch halfway out on some metal. Having had to return to the bomb to deal with that, he then had to risk his life again when the surfacing (17) became caught on wire netting. He looked in the fuze pocket. 'CAN'T SEE IF POWDER WORKED' his notes record, soon followed by 'STRIKER HAD MOVED ACROSS POWDER HAD NOT SET'. The hooked Zus had moved but not fired, and in the end he had to pull out the fuze. It was only by conducting tests on real bombs, rather than in the safety of the laboratory, that fatal flaws such as these could be uncovered, and lives not thrown away.

Hudson had had a narrow squeak on his very first day in his new job. The idea of using acid to cut into bombs had been abandoned in late 1940 after the UXBC had decided that mechanical methods were safer since they were easier to control. The research had been re-started, however, once the Services had realised that it might prove of help in cases where a bomb might be inaccessible to machinery. With Hurst and J. Gray, who headed the branch of the DSR concerned with research and development, Hudson therefore started his time in Richmond Park with tests on bombs filled with powdered explosive.

However, as aqua regia – a mixture of nitric and hydrochloric acid – began to course into one of the missiles yellow material started to erupt violently from it. Hudson and the rest thought it prudent to retire to the cow sheds, and twenty minutes later the bomb exploded.

Gough was furious. There had almost been another catastrophe, and he placed the blame at the door of the Woolwich scientists. Both Ferguson and the CSRD himself, Brigadier Macnair, had said that acid would not react with explosives. They in turn claimed that their advice had only been sought in respect of solid TNT. The spat grew worse when Captain Llewellyn, the head of naval bomb disposal, seized the opportunity to suggest once more that the MOS team were mere amateurs. He considered that experiments should be left to 'the usual channels, i.e. CSRD and Superintendent of Explosives, preferably co-ordinated by the Ordnance Board'. Meanwhile Stevens began to lure away Gough's contacts to do separate tests with acid for the RAF.

The DSR could see where this was heading. Gough's deputy,

E. T. Paris, thought 'it would be fatal to place all experimental
work with OB. This would mean that BD items would have to
wait their turn, and I believe it would virtually bring the work to
a stop.' Once again, the Services were trying to assert their right
to run all aspects of bomb disposal.

The renewal of this internecine conflict was part of a wider,
ongoing battle between the military ministries and the MOS. A
rapid turnover at the top of the latter, with Herbert Morrison
being succeeded first by Sir Andrew Duncan, then by Lord
Beaverbrook, and then by Duncan again, all in the space of eighteen months, had made for instability. Control of research and
development had passed through several hands, depriving Gough
of a powerful master who could fight his corner. It had also
allowed the Services to press their attack. As the official historian
of the MOS, J. D. Scott, writes, 'Fundamental doubts about a
Ministry of Supply responsible for research were hard to get rid
of.

'As late as 1942 the President of the Ordnance Board – a vice-
admiral – gave these doubts their most pointed expression by
saying that "A Ministry whose principal concern is production
(the antithesis of development) stands between the Board and the
initiators of requirements in the field of battle."' The tension
between the various groups – the soldiers, the scientists and the
civil servants, some of them now imported from industry – is well
caught in *The Small Back Room*, Nigel Balchin's 1943 novel about
BD. During the war, Balchin was deputy scientific adviser to the
Army Council, although his real expertise was in marketing;
while at Rowntree in the thirties he had helped the company to
create the Black Magic selection of chocolates.

Nevertheless, despite the strife within Whitehall, in the country as a whole the UXB situation had eased considerably by late
summer. In August 1941 Gough felt able to agree to a down-
grading of the priority rating given to bomb disposal equipment,
and by September Hudson could be spared as an instructor at the
new Army BD school. This was based at Donington Park, the

Leicestershire motorsport circuit, which had been shut down for the duration of the war.

Another instructor there was Len Jeacocke, who recalled it as 'peaceful and quiet – marvellous after London'. By the winter the course had moved to the School of Military Engineering at Ripon. One morning there in early December, Hudson was given some startling news: 'My batman brought in a cup of tea – it was a very cold day – and said that he had just heard on the wireless that the Japanese had bombed Pearl Harbor and that the Americans were in the war.

'I remember a great feeling of relief because I knew instinctively that "our side" must now eventually win the war. By no means had that been a foregone conclusion before then.' For him personally it was to signal a new adventure, as within ten days he had been recalled to London, and within a month found himself bound for America.

'I had been given some labels for my luggage, Cunard Line, but no name of a vessel, of course. So I reported to Liverpool with four Other Ranks who were with us. We went down with our luggage onto a tender. This tender kept on going, and we said goodbye to Liverpool, and we thought we were looking for a nice big liner, and one of us said to a sailor: "When are we getting to the ship?" And he said: "What ship? This is it."

'We went over on a 1100-ton ship – it was ever such a small ship! It was one of a fleet of banana boats. They used to scuttle across the Atlantic very fast, at twenty-three knots. There was no convoy as we were told we were much quicker than any submarine. I was seasick all the way.'

Based at the Aberdeen Proving Ground, Maryland, with three other officers, Hudson was given five months to train Britain's new allies from scratch in the skills of bomb disposal. 'They got together about thirty officers who were going to be their people,' he remembered. 'They didn't know what might happen, whether the Japanese might bomb America. They just wanted to learn. We had taken over our technical instructions and some fuzes, and we taught by seminar. They were nice lads – I kept in touch with some of them afterwards.'

The contrast between the austerity prevailing in England and the bountiful resources of the United States was startling. 'It was fascinating to see America gird itself for war,' he recalled. 'They were putting up huge barracks – luxurious – about one every two days. When we started BD, if you wanted a new piece of equipment you had to put in a form in quintuplicate. They just sent a chap out to buy it! They had money!

'They took very well to the work. They liked our inventions, but they nearly always invented something simpler to use and cheaper. They were good at that sort of thing. The point was by then we had defined what was needed. When we had been developing it, it had had to be done by trial and error.'

16

CONDUCT UNBECOMING

It was not until the end of February 1942 that the hero of St Paul's was able to receive his George Cross from the hands of the King. Much had happened since Bob Davies had raised the spirits of the nation a year and a half earlier, but his name still resonated with the British public. In their coverage of the investiture, newspapers had only to mention 'the bomb which fell in the vicinity of St Paul's' for readers to know which of the thousands that had fallen during the Blitz was meant. So it was all the more shocking when, later that same week, Davies was placed under military arrest.

His downfall had been a long time coming. As the list of nearly thirty charges drawn up by prosecutors revealed, for almost two years Davies had been balancing precariously between respectability and disgrace. And what was more, the Army had known that full well but had evaded its duty so as to avoid a scandal that might damage its image and national morale.

From the moment that Davies had arrived in London in June 1940 as a newly commissioned BD officer, he had begun to commit what was described at his court martial in May 1942 as 'fraud and dishonesty of the gravest kind'. In July 1940 he had put his section to work on the sly, building a private air raid shelter for a well-to-do resident of Lancaster Gate. For this he had asked twenty-five pounds (now about a thousand pounds). The materials were plundered from civil defence stores and the barracks at Bunhill Row.

As outlined by the prosecuting officer, Davies's thefts had rapidly begun to grow in scale and recklessness. After a section driver, Eric Jones, had been killed in an accident in July, his effects had been brought to Davies by his batman for return to his family. The next time that the soldier-servant looked at them he noticed that Jones's wristwatch and fifteen shillings in cash were missing. Davies told him that he had borrowed the money.

The onset of the Blitz offered further opportunities for gain. Although taking money from civilians grateful for BD's work was forbidden, Davies had accepted a cheque for twenty-one pounds with the assurance that it would go into a kitty for the section. He had cashed it himself.

He had gone far further in October 1940 when he had taken the five hundred pounds from Charringtons after clearing the UXB at their offices. Two weeks later, he had gone back to them and asked for another £190, saying that he been ordered to pay the original cheque into a benevolent fund but had already spent part on tropical kit for his imminent posting to Cairo. In fact, these were lies.

When Davies returned from Egypt in July 1941, the court heard, he had quickly lapsed into his old ways. It was invited to consider sixteen specimen charges of cheques he had bounced, an act regarded as an offence for an officer. The trail stretched over six months and from Plymouth to London via Doncaster, Halifax and Ebbw Vale. The last two cheques had been written when he was staying at the Russell Hotel in Bloomsbury prior to the investiture itself. Again, the sums involved were substantial, totalling £250 (now about ten thousand pounds).

Evidence was also given by a soldier that while working on the St Paul's bomb, Davies had taken a parcel of women's underwear from a warehouse near by that had been hit. He had given it to his batman, telling him to post half to Davies's wife and half to another woman. This Davies denied in court. He admitted that while waiting for the burning gas main to be turned off his men had helped themselves to the clothes, but he had told them to put them back.

As the judges must have known, four of the squad had been arrested the day after the UXB had been removed and had subsequently been convicted of looting. Davies was nonetheless cleared of this charge, and also found not guilty of having stolen Jones's watch and money. He had already pleaded guilty to having improperly received the money from Charringtons, and he was now convicted of all the other dishonesty charges and thirteen of those involving cheques.

The Judge-Advocate, Major G. H. B. Streatfeild, KC, said that the court 'would not be human if they did not recall with admiration the gallantry of Captain Davies, the men serving under him, and others, who not long ago were instrumental in saving lives and property in London.'

'But,' he added, 'considerations such as these can play no part in trying this case. Rather do they increase one's profound regret that an officer with such a distinguished record should have to face such charges.' Davies was sentenced to be cashiered from the Army and to serve two years imprisonment, without hard labour. His term was then reduced to eighteen months, probably as a nod to his GC.

The public shaming of Davies was an acute embarrassment for the Army. What emerges from the records of the case is more embarrassing still. Davies's misdemeanours had not suddenly come to light just after he was presented with the GC in 1942. Aside from the looting investigation in September 1940, a statement about Jones's effects had been taken from an officer, Lieutenant Berry, that October. But by then news had come through of Davies's medal.

Although Davies was asked for his own explanation of events, including the theft of the timber for the shelter, the nettle was not grasped. Rather, the first winner of the GC was got out of the way, being sent off to Cairo. Only a few days before he went he pressed the brewery for the extra money, thinking no doubt that it would be harder to check up on him once he was abroad.

By the time he returned, the evidence later used at his court martial had been before senior officers for the best part of a year. It was, however, to be almost another eight months before any action was taken. The most probable explanation is that the Army had dithered, wanting at all costs to preserve the glow lent by St Paul's. One pointer to this is the surprising length of time Davies had to wait before receiving his decoration in February 1942, given his prominence and the fact that he had been back in Britain since the previous July.

Having had their hand forced by the spree of bad cheques, the generals had wondered whether it was more humiliating for the Army if Davies was disgraced before or after he had met the King, while hoping that the problem would go away. As the Army lawyer who reviewed his sentence noted, even once Davies was in custody there had been three months of paper-shuffling before he was brought to trial. By then things were less desperate on the Home Front and a new symbol of defiance had been found: Malta, GC.

The excuse presented for Davies's behaviour by his defending counsel was that he had got into debt while trying to repay hospitality after he had become a national celebrity. It was understandable: the GC had gone to his head; he had started to live a little extravagantly. He would not be the first idol shown to have feet of clay. Davies himself claimed in court that he had agreed to build the shelter out of his 'generosity of heart'.

If his misdeeds had been isolated incidents, this might be credible as a motive for his actions. Yet they were part of a broader pattern that reached back not just to the months before his rise to fame but deep into the decades before that. Davies's life had been built on economies with the truth. From that first fib about his age in order to enlist in the Canadian Army, too many steps along the way had been marked by vagueness and inconstancy. Even the children of his second, post-war marriage were given few firm details about the first fifty years of his life. Again, he would not be the first man to want to appear bigger than he was. Nor would his psychology be of much concern if it did not chime with inconsistencies more important than mere dishonesty about money.

Those inconsistencies go to the heart of the story of the St Paul's UXB itself. The version propagated by the media, and by Davies's citation for the GC, was that the cathedral had been threatened by a huge time bomb that might explode 'at any moment', and 'in order to shield his men from further danger' Davies had driven it to Hackney Marshes and 'personally carried out its disposal'. But that is not the course of events supported by the evidence.

In the days following their exploit, while it was still becoming a public sensation, the squad gave a series of unguarded interviews to the popular press. Readers learned that it was the NCO in charge of the transport section, Lance Corporal Bert Leigh, who 'drove the lorry out to the Marshes, alone – preceded by motor-cycles and followed by ourselves'. Davies is conspicuous by his absence from the section's remarks about who had done what. Only in *The Times*, the newspaper of the Establishment, was the officer credited with having taken on the responsibility.

Examined objectively, the notion that Davies could all by himself have unloaded and exploded a bomb weighing a ton becomes implausible. It is much more likely that it was done as a group, and George Wyllie's wife, for one, remembered having first seen her future husband while he was in the convoy that had sped through Hackney. That Leigh was given the BEM when the awards were announced suggests that he had played a central role in what had really happened.

It is noticeable that it was the press, too, which first made the assumption that the UXB had had a time-delay fuze and that the squad and cathedral had been in constant danger. This was perhaps natural: the (17) was the new menace, at the forefront of everyone's mind. Yet the sappers themselves never mention the fuze in their comments, nor is there a record of it in the Company War Diary.

For whatever the press chose to think, what undermines the dramatic tale of the giant time bomb are hard statistics. The plain truth is that German armourers did not put (17) fuzes into the type of UXB found at St Paul's. A thousand-kilogram Hermann was designed to detonate on impact; anything else was a waste of a metric tonne of steel and explosive. If the objective was to disrupt and to deny territory to an enemy by using a time fuze, that could be accomplished just as well with a much smaller bomb.

The records speak for themselves. Soon after the St Paul's incident, figures on bomb size and fuze type began to be gathered systematically by BD. They show, for instance, that in the month after Davies's deed ten thousand-kilogram UXBs were dealt with. None had (17)s. Those UXBs that did were the much more common 250-kilogram, 40 per cent of which were found to have clockwork fuzes.

The pattern continued for the remainder of the Blitz. Between late January and early March 1941 a further thirty-three Hermanns were uncovered. Again, none had a (17). In fact, the majority had (28)s, the short-delay fuze set to cause maximum damage after penetrating a reinforced target. Variants of it were used in most bombs over a thousand kilograms, such as the enormous one uncovered by Wyllie at Mount Pleasant and those encountered at the Welsh ports by Archer and Talbot. This would be consistent with the St Paul's Hermann having originally been aimed at the docks.

Mistakes can, of course, be made by armourers. It is possible that the wrong fuze was put in Davies's bomb. Possible but, on the evidence, highly unlikely. A study in the seventies by the Ministry of Defence's expert found three reports of (17)s in the 261 Hermanns dealt with between late 1941 and the end of the war.

Taken with the earlier figures, this suggests that about three in three hundred – or 1 per cent – of thousand-kilogram UXBs were found with time fuzes.

In addition, those three hundred UXBs were just the bombs that had not exploded as intended, and so represent only about 10 per cent of all Hermanns dropped. In other words, the chances of a (17) being absent-mindedly inserted and then being discovered at St Paul's was approximately one in a thousand. On the facts, let alone on the balance of probabilities, everything points to that fuze not having been a ticking (17) on the verge of exploding, but a dud (28).

That is not to say that anyone in the squad lied about what had taken place at St Paul's. Despite Wyllie's claim that they had dealt with 'hundreds' of bombs, they were actually very inexperienced, and he had only been on the job a few days. For instance, later in the war he would not have clanged the UXB – in his phrase, 'tapped it with my crowbar to make sure' – as it might have freed a stopped clock. If he did uncover the fuze, he may not have understood the significance of its number.

It was likely that only Davies, as the officer, knew that, and then only if it was discernible. As the fuze was not discharged on the spot it was later presumed that it had been a Zus-protected (17), giving Davies no alternative but to move the bomb. In actual fact, it is more probable that its location meant that he could take no chances even with an impact fuze, which might turn out to be booby-trapped. The priority was always to get the weapon as far away as soon as possible.

What would explain these inconsistencies is an alternative history of the events at St Paul's, one that would suggest that they were – to coin a phrase – sexed up. Excitement at the symbolic salvation of the cathedral led to misunderstandings and over-eager reporting. A daring officer and a time bomb, with its extra frisson of danger, made for a better story. So good a story, in fact, that it was just what senior officers and the government were looking for.

'The valuable psychological effect of immediate awards for gallantry in "blitzed" towns has always been recognised,' ran one
MOHS report of the period. London was just such a town then,
and it needed the lift that the first George Cross would provide.
Within a couple of days the news began to snowball and then it
was all over the papers. No one seems to have been allowed to
seek Davies's own account of the story, which was confined to the
official version in *The Times*.

The first that the sappers knew of the GC was when they
heard about it on the radio. Until then, all the acclaim had been
a bit of a lark, a bright spot amid the dark uncertainties of their
work. Now it was too late to go back. If anyone had their doubts
they kept quiet or were ordered to do so. There had already been
too much trouble with the looting and the shelter, and that fire in
the office which had destroyed all their records.

Nor was Davies one to turn down such an opportunity. So,
perhaps also under pressure from on high, he played along. His
courage is unquestionable – he risked his life on many other
occasions and for that entirely merited his decoration – but he
would not be alone in being both a brave man and one who
muddied the waters of truth.

Davies was the one person who knew the whole story and it
was he who must have been the most uncomfortable with how it
was spun by the Army and the press. It is certainly a pity that no
detailed record survives of the version of events presented to the
GC Committee. Many other of their deliberations – such as about
Archer's medal – were selected for preservation in the Army's
archives, but those for St Paul's were not, even though they were
the first to concern servicemen.

From their subsequent behaviour, it might fairly be argued that
both Davies and Wyllie found the medals they had been awarded
a burden. Following his decoration in June 1941 Wyllie grew
progressively more disenchanted with BD.

Having been promoted to corporal, he was asked to instruct
conscientious objectors in the work: 'I told them that if they

were not good enough to fight at the front, then I wasn't going to train them.' A series of transfers followed. His private life became messy; one young officer was startled to be woken in his bunk bed by the sound from that below of Wyllie 'entertaining a young lady'. In April 1943 Wyllie used his old football injury to get himself a discharge from the Army. He worked first at a battery factory and then, for thirty years after the war, at the Ford foundry in Dagenham.

He had little to do in that time with BD reunions, and in 1984 announced that he was selling his GC. He was not well off, lived alone and felt disappointed by modern Britain. 'I don't think now the country is the same one that I knew when I got the medal,' he said. It was bought by a merchant bank and presented to St Paul's. George Wyllie died three years later, aged seventy-eight.

The decision by a living holder of the GC to sell his medal was described at the time by the Victoria Cross and George Cross Association as 'almost unprecedented', so highly was the honour esteemed by those who had won it. Virtually the only other person to have sold their medal in the same fashion was the disgraced Bob Davies.

After his release from prison in 1944, Davies had understandably dropped from sight. He also was prone to complicated relationships that stemmed from a fondness for women, and his stories had now been found out. He was divorced by his wife and in 1949 immigrated to Australia. There he made a new life, raising a second family and setting up a building business. He died in New South Wales in 1975.

His George Cross had been sold in the years immediately after the war. Perhaps it carried too many unhappy memories. At the least, the aura of the medal seems to have exacerbated his weakness for falsehoods and emboldened him to commit worse wrongs. Davies had been living a double life and it became second nature to him.

That the Army turned a blind eye to his conduct so as not to tarnish the golden legend of St Paul's would appear incontrovertible. Some questions could also be asked about the part it played in creating and perpetuating that myth. There is a case to be made that nothing can be gained from raking up the past, but it is not the case for the truth. That comes from hard questions, not easy answers.

17

Y FUZE

It was the middle of the night when the fuze was brought in. John Hudson was on duty and he knew immediately that BD had a crisis on its hands. He telephoned his air and naval counterparts and got Bob Hurst out of bed. Within an hour they were sitting around a table taking apart a fuze unlike any they had ever seen.

From the outside it looked just like a conventional (25B) impact device. Yet the first thing they saw when it was opened up was not condensers but two small batteries. The upper part of the fuze was revealed to resemble a (17A), though it served on impact only to arm the portion below. In this were four anti-handling switches. One was a trembler like that in the (50). The other three were more sinister: mercury tilt switches embedded in wax and set at right angles to each other. Any movement of the mercury would complete the circuit.

The evil genius of the fuze lurked at its base. This was tapered like a cone, around which sat a split ring of wire. It rested on a bevelled lip above the gaine, so that when the fuze was tugged upwards the ring was held by this edge and the switches tripped. The only way to withdraw it was by using a twisting motion, which would also detonate the bomb.

This was not a fuze that could be countered in any manner known to BD. There were no moving parts to jam or to magnetise. It was like a (50), in that once armed it waited for its victim, but the energy in a (50) eventually leaked out or could be discharged. This was powered by sealed batteries of a type that they knew might last for up to a year.

A (50) could be avoided. The great fear had always been that the Germans would begin deliberately to misnumber their fuzes, but until this time one glance at the markings on the head had been sufficient to tell a BD officer with what type he was dealing. This was different. This would look like a (25) or a (17) and no one would know it was not until it seemed safe to extract. Then it would be too late.

This was a wolf in sheep's clothing, one that would not strike until touched, and nothing less than a calculated attempt to kill bomb disposers. And until Hudson and his team could work out how to beat it, no fuze in any UXB could be trusted. All BD work would have to cease indefinitely.

Until that night, Hudson had enjoyed six months of relative calm at BD headquarters. He had returned from America in early June 1942 to find that General Taylor had been succeeded by his deputy, Brigadier Harold Bateman, but the handover had not resulted in any far-reaching changes. The following month, Bateman recommended him for promotion to major as 'possessing outstanding ability, drive and technical qualifications'.

Hudson was pleased with his new responsibilities as Deputy Assistant Director of BD (Technical) – 'a bit more pay, of course!' – although he missed the company of Gretta and his two young sons, Colin and Dick. They were by now staying with his father in Chapel-en-le-Frith and, aside from the occasional bit of leave, he had been apart from them since the first months of the war.

Gretta tried to sustain him with cheery letters that gave him a sense of home life – 'Colin says the teacher puts a kiss on his sums if he gets them wrong' – but sometimes her loneliness surfaced in them. 'I wonder what you are doing this evening,' she wrote one Saturday evening, 'perhaps a good old fashioned Sat pm with a book.

'How I long for those days again when we have those delight-ful weekends together again, the days given up to the boys and the evenings to ourselves . . . It's a bitter disappointment to me also that you can't share the development of our boys with me, I always feel that I enjoy them more wholeheartedly when you are with me.'

There was plenty, however, with which Hudson could busy himself in London. Following the end of the Blitz and the most pressing demands on BD, the R&D sub-committee had gone into abeyance in June 1942 and its functions had been subsumed by the UXBC itself. Development had nonetheless continued of simpler and better equipment, notably of electric trepanning machines.

The entry of America, and of Japan, into the war posed new questions, while the passage of time had varied old problems. It was discovered, for example, that after a few months the fillings in bombs became unstable. The ammonium nitrate in aluminised powder released ammonia that could be lethal in the confines of a shaft, and contact with the powder turned the plug of TNT in the nose of a buried Hermann into nitroglycerine.

Fuzes that retained their charge for many months were found, but there was only one example of a UXB exploding without warning after such a time. The afternoon of 6 June 1942 had been sunny in London, and the air remained pleasantly warm until late into the evening. Couples took the opportunity to go for a walk, and in Gurney Street, close to the Elephant and Castle, the children had been allowed to stay out a little longer, playing cricket and riding their bicycles.

John Hudson had returned from America the day before, and was in his office in Westminster when he heard a 'sudden, loud clap of thunder' from across the Thames. Young Danny Slattery witnessed the explosion from much closer: 'Everybody dropped everything. We were playing cricket with an old floorboard and a tennis ball, and we all looked up. Bricks were coming down and there were gas pipes, water pipes, all blowing everywhere. We all ran.'

A bomb which had lain hidden for more than a year had detonated, bringing down a block of flats and demolishing five other houses. Peter Vigus was walking close by when he saw 'a big girder – it must have been eight or nine feet long – which went right over the top of the first lot of buildings and into Lion Street . . . I started running towards where the bomb had gone off. Everyone did that sort of thing, to see if you could give a bit of help.'

Two hundred tons of masonry had collapsed into a crater twenty feet deep, and as the rescue and demolition squads sifted through the wreckage they found twelve corpses. Several other people were brought out alive. Then, as the road itself began to be cleared of a vast mass of debris, a heart-rending discovery was made: the crushed bodies of six little children who had been at play in the street. The body of a seventeen-year-old girl who had been sitting on a doorstep of her home, perhaps catching the last rays of the sun, was found later.

Fourteen-year-old Iris Ward was traumatised for years afterwards by her narrow escape from death. 'It was just as if someone had opened a great big oven,' she recalled. 'The heat rolled, literally rolled up the street . . . I must have shut my eyes. I felt myself going up and a lamp post bringing me down. When I opened my eyes, everything was dark. I could feel something on top of me. It was either a little girl or a little boy, and I thought, I can't have this, and I pushed it off me.'

The cause of the explosion, which injured sixty people and left two hundred homeless, was never ascertained. One theory was that a time fuze, perhaps dropped in the 10 May raid of the year before, when the area had been hit by bombs, had been jolted into life by vibrations from the Underground or from a house that was being pulled down.

It was a measure of how much more secure the government felt about the threat from bombing that the press was allowed to give detailed coverage of the incident, and the tragedy led to renewed efforts to locate other UXBs from the Blitz that had not been accounted for.

The Luftwaffe added few new bombs to these in 1942. Its fleet in the West was now so small that the occasional raids it mounted averaged only about twenty aircraft, and where it did do damage it was by bombing ill-defended targets such as Bath, Exeter and Canterbury in the Baedeker raids of the spring and summer. Towards the end of the year the attacks intensified, however, and that directed at London's docks on 17 January 1943 was the heaviest experienced for some twenty months.

The next morning, Major William Parker was passing Lord's cricket ground, which had been requisitioned by the RAF, when he noticed some earth and rubble strewn at the foot of a wall. His suspicions that a bomb had fallen were confirmed when a five-hundred-kilogram UXB was found to have penetrated a tunnel on the Bakerloo line, which runs under the pitch. Since it had come to rest under the track it was given a Category A rating, and Captain Frank Carlile and a section from 5BD Company were sent to deal with it.

The only slightly unusual feature to the (25) fuze uncovered in the bomb was the letter 'Y' also marked on the boss. This did not seem to have any great significance – it might indicate a date or place of manufacture – and so Carlile began to use the liquid discharger, as was now standard. Only about a quarter of the usual volume of liquid would go in, however, even after two attempts. Nonetheless, this was not so strange as to cause alarm, and Parker agreed that the bomb could safely be hauled to the surface by a crane.

It was taken to the UXB cemetery on Hampstead Heath, where Carlile tried a fuze extractor on the bomb, but still without success. Intrigued, frustrated and satisfied that the fuze was inert, he resorted to recovering it by brute force. He hacked away with a chisel and then a hammer before finally levering it out with a crowbar. Only once the fuze was in his hands did he realise the danger he had been in. Within a few hours the Y fuze, as it was christened, was on Hudson's desk.

While Hudson and Hurst began to study it, other reports were coming in of suspect UXBs. One had fallen in a Battersea warehouse full of new machine tools from America. After tearing through steel roof girders and packing cases, it had come to a stop resting almost vertically underneath the bed plate of a very large lathe.

The first problem was to establish whether either of the two fuzes in this five-hundred-kilogram bomb was a Y. Anticipating such an emergency, some experiments had previously been made with radiography, establishing that X-rays could be used to view the contents of a fuze without having to touch it. This was given the deliberately vague name of 'field photography' so as to conceal the essence of the technique from the Germans. After twelve hours of work in the warehouse Dr John Dawson of the MOS and Captain Alwyn Waters, a BD officer, were able to take an image that confirmed that, despite its camouflage, one of the fuzes was indeed a deadly impostor.

There was no way of extracting it safely, not least because the casing had been much distorted by its passage through the building. Nevertheless, it was imperative that the tools were spared damage, and that a large flour mill next door, which had been shut down due to the vibrations its machinery was emitting, was re-opened as soon as possible. The UXB was therefore designated Category A, and allocated to Major Cyril Martin.

Martin was another of BD's veterans of the Great War, during which he had, at eighteen, won the Military Cross for extinguishing a blaze in an ammunition store. The son of a Derbyshire priest, he had worked as an electrical engineer between the wars, but had returned to the RE in 1940 and, in his mid-forties, had volunteered for bomb disposal. His only son had been killed the following year when the corvette in which he was serving was sunk by a U-boat off Gibraltar.

Martin and his assistant, Lieutenant Ralph Deans, had very little to guide them but decided that, since the bomb could not be moved and the fuze could not be extracted, the only way to make it safe would be to steam out the explosive. By midday on 20 January they had managed to unscrew the base plate at the back, only to discover a new complication. The entire filling was solid TNT. This would have to be emulsified by the application of steam at a high temperature, but if that was done by the usual remote process there was a substantial risk that the force or the heat of the jet would trigger the fuze.

The only solution would be to do it by hand, applying the nozzle at close range and for just long enough to soften small areas of the TNT that could then be scraped away with a trowel. But that would take much longer, and place them all the while in extreme danger.

There was no alternative. Gritting their teeth, Martin and Deans dug out a space beneath the lathe so that they had more room in which to work. The bomb was then secured upright as best they could, before both men got into the cramped pit under it and turned on the steam. The working conditions in the hole were soon appalling. It was like being in a Turkish bath a dozen feet below ground. They lay on their backs as they probed, with hot explosive dripping on to them and cold water soaking their clothes. Every time they breathed, they could feel particles of TNT in the air.

On they laboured without a break, all afternoon and then right through the night, wearily chipping away at the explosive and conscious that any slip could jar the trembler switches in the fuze. Other bombs were falling nearby. Finally, after spending almost twenty-four hours under the UXB, they succeeded in scooping out all of its filling. The case and the Y fuze were then disposed of.

Hudson believed that BD was a 'dangerous job which called for cool efficiency and technical knowledge rather than heroics'. Frequently, as Deans and Martin had shown, all three went hand in hand, and in March they were awarded the GM and the GC respectively. By now convinced of the validity of such decorations for bomb disposal, Sir Alec Hardinge commented of Cyril Martin's recommendation: 'What a marvellous story it is!' Martin's wife Jessie was not so sure about all the excitement.

'My husband's absorbing interest in life,' she told the press, 'apart from his family, is his work. But he is a very modest man, and will not be at all pleased at the publicity about this award.'

Meanwhile, the boffins were beginning to make progress. It was evident that the sole weak spot of the Y fuze was its power supply, and it was against this that any attack must be concentrated. Ordinarily, batteries were invulnerable to methods such as immersing them in a conducting liquid, and the only hope was to find some way of super-cooling those in the Y fuze so much that they died. Then Bob Hurst remembered that he had some liquid oxygen in his office.

Hudson's memory of the moment they huddled around the fuze was vivid: 'We put a lead and a galvanometer on one of the batteries and put some liquid oxygen poured into a basin, which was all we'd got. We dropped the battery in – and instantly, almost instantly, the output from the battery fell to zero. So we knew if we froze the battery enough it would be dead. Then we had to find out if we could freeze the battery in the middle of a bomb.'

They wired up a measuring device to a fuze inside a bomb casing and began to drip liquid oxygen on to its head. 'We dripped it on for quite a long time before anything happened. This was all done in a cellar. We must have created quite an explosive atmosphere with the liquid oxygen . . . We were very careful not to create a spark.

'About ten or twelve minutes passed. Then, quite suddenly, we noticed the meter starting to fall. It fell faster and faster until suddenly it went to zero.' That told them that if they applied enough liquid oxygen the battery in a bomb would eventually go dead, but they still had no way outside the laboratory of measuring exactly when.

'And then Bob Hurst noticed that as we started dripping liquid oxygen it started to make a ring of hoar frost on the bomb, taking moisture out of the atmosphere. This ring gradually got bigger and bigger. I think it must have been Bob who had the wit to say that the ring was that big when the battery went dead. So if we waited until it was that big, we should be safe.'

By lowering the temperature of the batteries to below −30°C their internal resistance became so high that they would not fire the igniters in the fuze. But this technique had only been tried in controlled conditions, and on just the single specimen. There was no more time to do further experiments. The only way of proving that it worked in the field would be to test it on a live Y fuze.

Despite now being a senior staff officer, Hudson volunteered to conduct the trial himself. 'I thought that the right approach was for the chap who developed new ways of dealing with a fuze to try it out,' he argued. 'If I'd given instructions, having thought up a wheeze, and the first chap who tried it had been killed, then what would I have felt like?'

Not far from the warehouse with the machine tools, another Y fuze had been identified. This had fallen close to the Albert Bridge and the Embankment. No traffic could get across the Thames or enter an important flour works. Hudson arrived at the site on the sunny Sunday afternoon of 24 January.

Although he had confidence in the freezing technique, there was a real danger that when the liquid oxygen was used it might contract the bomb casing and the fuze pocket unequally, provoking a sudden fracture that would cause the fuze to function. He had therefore arranged to give a step-by-step commentary of all he did by field telephone, so that if anything went wrong it would help his successor avoid the same mistake.

Hudson had brought with him a packet of plasticine. One problem that he had yet to solve was how to prevent the liquid oxygen from dispersing too rapidly. Moulding the clay with his fingers, he built a crude cup around the rim of the fuze head, then he poured the liquid into this.

The team had worked out that when the frost formed a ring with a radius of one foot the battery could be considered safe. The simplest way of measuring this that they could devise was to stick wet pieces of cotton wool on to the metal casing at one-inch intervals from the fuze. Hudson had to wait until the last one had frozen solid.

For two hours he monitored the white strips, peering through the dense fog thrown up by the evaporating oxygen as he topped up the pool in the plasticine reservoir. There was a nasty moment when, as he had feared, the casing cracked, but it failed to trigger the fuze. Finally the circle of frost did its work.

Hudson reckoned that he now had a maximum of twenty minutes before the batteries began to warm up. In order to avoid having to twist the fuze out by hand he planned to use a line to give it a straight pull from a safe distance. When he retired to cover, however, he found that a brigadier and two colonels had come to watch the operation. He thought it diplomatic to offer them the honour of despatching the fuze, but in their eagerness they pulled too hard on the cord. It broke and they fell over in a heap. Hudson had no choice but to return to the bomb.

Aware that the frost was thawing, and with the light failing, he began to prise open the icy fuze pocket with a jemmy. The boss was countersunk level with the case itself and purchase was minimal. As the seconds ticked away he attacked it more and more furiously, wrenching and turning what he could grasp. Eventually the frozen fuze yielded. He looked at his watch. Twenty-three minutes had passed since the cotton had frosted over. Hudson unscrewed the gaine and slipped it into his pocket. The Y fuze was beaten.

It had taken just a week to free BD from its state of paralysis. Work could now return to normal, but there remained half a dozen other Y fuzes to deal with in London. Hudson tackled the first, to confirm to his satisfaction that the immunising technique could be used by the sections. Some of those that remained to be dealt with by his fellow officers proved trickier.

One was found about two weeks later, about ten feet under a side street off the Old Kent Road. Alwyn Waters began to lever it out, only for the head to break off, leaving the trunk jammed deep in the fuze pocket. There was nothing for him to grip, and in a few minutes the batteries would be lethal once more.

Martin was with him and decided to handle the situation himself. Slowly he began to dribble liquid oxygen on to the exposed fuze, chipping away pieces of it as they froze. Bit by bit he cut through the electrical components until he was able to break the contacts with the mercury switches. Had he sliced just a little too deeply he would have struck the percussion cap. His original plan had been for him to describe to Waters what he was doing through a microphone, but soon he became so focused on each blow with the chisel that he lapsed into silence.

Two days later Martin went to a house where another Y fuze was lying at the bottom of a deep and ill-ventilated shaft. It was being frozen by Frank Carlile, the officer who had brought in the first example, and he had just begun to extract this one when he was engulfed in flames. The highly oxygenated atmosphere had spontaneously ignited a substance in the shaft, which then started to belch fire. Although both his clothes and the ladder were ablaze, Carlile managed to climb to the surface, where Martin and another officer put out the flames. Despite the risks, Martin went down to the bomb and made it safe. Carlile was terribly burned, but by August he was back on duty. Later that month he was killed by a fuze that exploded at a bomb dump at Horsham. He was forty-two.

Bombs containing Y fuzes continued to be dropped on Britain for more than another year, but even though Herbert Rühlemann's team continued to improve them – for instance introducing a battery that froze at a lower temperature still – they never again proved a serious threat to BD. Nor did the Germans ever make another such attempt to deceive their opponents with the sole purpose of killing them. Nevertheless, the advent of the fuze had sparked a debate that had run for much of the war.

There were many in BD who were convinced that each advance by the Germans in fuze technology was evidence not of Teutonic ingenuity but of lax British security. No sooner had the Crabtree discharger been devised than fuzes were rewired to explode when it was used on them. Once (17)s began to be extracted while still ticking, the Zus appeared. The invention of the Stevens Stopper was met by a fuze immune to it – the Y.

In the early days of the Blitz the chief culprit was felt to be the press, which gave away many details of BD's work while fêting their exploits. John Strachey met two officers who were dealing with delayed-action fuzes in Chelsea: 'It was just after Lieutenant Davies had taken the ton bomb out of the foundations of St Paul's, and the newspapers were full of the squad's prowess. The officers deplored this. They argued that if the Germans were given the impression that we were dealing successfully with their present types of D.A. bombs, they would invent new booby traps to catch those who had to remove their detonators.'

In November 1940 Herbert Gough had written in similar vein to General Taylor, pointing out that a photograph in the *Daily Telegraph* of a bomb being removed clearly showed a fuze key. 'I regard any publicity of this kind as highly undesirable,' considered Gough, 'and as likely to provide the enemy with information about the methods and apparatus we have.' Thereafter, BD's activities were covered by the newspapers with far more discretion, although during the Blitz that did not prevent them from being described in detail by publications in neutral countries, notably America.

Given the paucity of intelligence that the Germans were able to gather within wartime Britain, most of Rheinmetall's fuze development in fact owed far less to careless talk than to their anticipation of likely countermeasures. Nonetheless, Rühlemann and his team did profit from captured British documents and equipment, such as in North Africa. They were even up to date with the fate of the Y fuze.

When interrogated at the war's end, the Rheinmetall technicians reminisced about it as one of their most pleasing creations, 'an excellent one of its type'. They also revealed that they knew from a Luftwaffe briefing paper that 'liquid air was being used to neutralise this fuze', and were frustrated by being unable to think of a way to beat this.

Yet information flowed both ways. By 1942 BD had seen instructions by the Luftwaffe's Quartermaster General (Air Equipment) that all fuzes unsuitable for low-level attacks should have their tops painted yellow, usefully marking them to bomb disposers as (28)s or (35)s even if their numbers had been torn off on impact. The RAF had also by then had the benefit of reading orders to German disposal teams about handling British fuzes, allowing those to be varied. And, following the capture of a German pilot, the UXBC was able to declare in 1943 that it had been 'learnt that these Y fuzes were used on the express instructions of the highest German authorities for what was intended as a direct blow at the morale and efficiency of our bomb disposal squads'.

Many of the German documents in the BD archives bear Hudson's signature, as proof that they had been assimilated into his bulletins to the sections. Shortly after his battle with the Y fuze, Gretta had written to him, flushed with both worry and pride: 'John my very own beloved,' she began, 'this has been one of the serious days of my life – I'd no idea until I read your letter this morning that you yourself had been in actual danger last week.

'I knew that you had a great problem to solve, but never dreamed that you were actually in contact with the horrid thing – thought it was just research work etc. My dear, I love you more than ever & only regret that I can't take your dear head in my arms & tell you so . . . I've really felt topsy-turvy since reading your letter.'

By Easter, news had come through that John had been awarded the George Medal for his display of courage at the Albert Bridge. 'I was so excited to get your letter this morning,' wrote Gretta. 'I feel so very thrilled about everything, especially the incredible story about the King. What a lot you'll have to tell us next weekend, I do hope you'll be able to get one night away, just time to pop down . . .'

Her letters, and the experiences they speak of, stand for those of every couple parted by the war. Hudson was able to spend that weekend with them, but Gretta longed for the day when there would be no more goodbyes: 'Colin didn't cry this morning but he walked back into the house and had great difficulty in keeping his lips from trembling when he asked me "How many days will it be before Daddy comes again?"'

Her own need was no less acute. 'It's impossible to put into writing all my longing for you that has been overwhelming these last few days . . . I do love you so very dearly & sincerely John, my heart is very lonely without you.'

18

BUTTERFLIES

George Shoebottom was a good sergeant. At least, so thought his fellow Yorkshireman Wally Fielding, who served under him in East Anglia from the start of 1941. Shoebottom had seen them through some nasty moments, such as when a bomb had gone through a house at Saxmundham, destroying a bed and the person sleeping in it. There had been lighter ones as well, such as when a UXB had bounced on a road at Lowestoft and come down in a room where the Home Guard was meeting. They had been called out to deal with it and afterwards Shoebottom had gone to sleep on a trestle table there; for a bit of fun, some of the lads had tied him to it.

They had laughed about that when he woke up. They had laughed when Lord Bowater had solemnly given them a pound after they had spent days defuzing bombs at an agricultural school he owned – a whole shilling each! Then they had been told that the Army was to give them an extra shilling a day for food. There had been a bit of a debate about that. Some of the men wanted to send it their families, but the sergeant said that it had to go on food. They had fourteen pounds in the kitty, which was a fortune. Rationing was starting to make things scarce, so Shoebottom had bought a huge jar of pickled onions.

On a Sunday afternoon in the late August of 1941 they had
been out in Essex. 'What I liked about bomb disposal was being
in the country,' recalled Fielding, who in peacetime had worked
in one of Sheffield's leather factories. They had heard the sound
of a sharp bang, and George Shoebottom and Lance Corporal
Lewis Farquharson were dead. What had killed them was one of
the most insidious weapons in the fuze war, and perhaps its best
kept secret.

The *Splitterbombe* SD2 was, owing to its shape, better known as
the butterfly bomb. One of the first anti-personnel cluster bombs,
it was dropped from containers that held two dozen of the coffee-
jar-sized devices. As they fell, their cases sprang open, forming a
pair of rudimentary hinged wings that rotated in flight like a
sycamore seed. This armed the device by twisting out a spindle as
it drifted to earth.

The tactical intent behind its use was to cause deaths and
injuries among those rushing to put out fires started by the incendiaries with which the *Splitterbombe* was dropped, so hindering
efforts to fight the flames. The butterflies weighed just two kilos,
and their disc-like wings would become caught up on telephone
wires, roof gutters and tree branches, waiting to be triggered.

Although some had impact or delay fuzes, the most common
type was armed by concussion, meaning that any subsequent
movement of it – including strong vibration – could set it off. It
held only 225 grams of cast TNT, topped with bitumen, but the
thick steel of its case was scored around and would splinter into
fragments lethal at thirty yards and capable of wounding at a hundred.

The first raid with butterflies came in the autumn of 1940,
when a handful were dropped near Ipswich. One was recovered
when an alert sergeant, Charles Cann, noticed that the arming
spindle had not unscrewed completely, allowing him to make it
safe. Usually they were painted dark green, but in the harvest
months this colour scheme changed to yellow, better to conceal
them amid the ripening corn, and several more attacks were made

in the summer of 1941 when Shoebottom was killed. They also fell occasionally on aerodromes in eastern England – the aim being to harm flight crew – but only in the invasion of Russia were they used en masse, and to great effect, against concentrations of troops.

Since the attacks were neither widespread nor frequent, warnings about them were confined to the areas where they landed, and by 1943 they had largely been forgotten about. Then, on the night of 3 March, some 1800 butterfly bombs fell around London and the South-East, principally in rural areas. Several landed on Epsom racecourse, and John Hudson's former commanding officer at Halifax, Ray Bingham, was sent to clear them. He placed one-pound slabs of gun cotton on a few specimens and then retreated to the safety of an armoured car, only to be killed when the blast blew a butterfly into the vehicle, where it detonated.

For the first time, announcements were made locally in newspapers and on the radio about the bombs, and posters depicting them were sent to schools and police stations. Determined efforts were made to find and remove those which had landed in ditches and hedgerows, or arrangements were made to let them lie until standing crops had been harvested. The great fear was that butterflies would be used against the capital, but though there were inter-departmental discussions about the merits of greater public awareness little more had been done by the time the summer arrived.

BD had had a quiet few months, at least in Britain. As the war had begun to turn several companies had been posted overseas, mainly to North Africa and the Middle East. As well as dealing with unfamiliar Italian and Japanese bombs and mines, they had had to cooperate with RAF teams, who were supposed to deal with the increasing numbers of Allied UXBs found as the advance on both fronts gathered momentum in 1943.

In practice, however, there were too few Air Force specialists available to tackle the workload. In Algeria and Tunisia the RAF had sufficient numbers only to cover aerodromes, while the RE had to take on the rest. Yet such was the Air Ministry's obsession with security that for months it refused to release the necessary technical information about its fuzes to Army personnel.

This led to a lengthy feud between it and the War Office, which became furious about casualties caused to BD teams by British UXBs. The War Office was also mindful that any seaborne invasion of Italy or Normandy would be preceded by heavy aerial bombardment. By the autumn, it was reporting that 87 per cent of all unexploded bombs being dealt with by the Army in the Mediterranean theatre were British, but it was not until the end of the year that the Air Ministry agreed to allow it access to details of all RAF fuzes then in service.

At home, meanwhile, the opportunity was taken during the summer to send several companies on training courses. Then in the early hours of Whit Monday, 13 June, thirty German aircraft flew in from the east, over Grimsby and Cleethorpes. They scattered in their wake eighteen tonnes of bombs, including six thousand incendiaries and more than two thousand butterflies.

Many of the incendiaries landed near the Fish Docks, where a large fire started. Some of the bombs were fitted with small explosives under their tail to discourage interference, so when more than one landed in a house firemen could find themselves cut off from a blaze in an upstairs room by a lethal incendiary on the ground floor. The Great Grimsby Coal, Salt & Tanning Company's rooms were soon burning furiously, and incendiaries also lit up the wooden Fish Market. Trawlers had to be moved away from the quay while fire-fighting tugs were deployed, but by a quarter to four in the morning more than 330 fires had been reported in the town. Weelsby Old Hall Hospital became an inferno; its patients were transferred to Grimsby General. The tram lines were brought down by high explosive, and the air filled with a choking mix of smoke and brick dust.

Rescuers did what they could. Two hundred naval ratings formed a bucket chain. Fred Dolphin, a sailor home on leave, climbed into the wreckage of a house through a hole where the chimney had been after hearing a woman call for help. She was pulled to safety, and after giving him a kiss ran off into the night.

Worse was to come. As both civilians and ARP workers began to move into the open, to check their property or to stare at the destruction, they began to stumble over the butterflies unleashed on Grimsby's streets. Hanging from bushes, littering the open spaces of Sidney Park or lodged in window boxes, they claimed more than thirty lives in the first hour after the raid as people picked up the unfamiliar objects or brushed against them in the dark. Five of the dead were policemen, four more wardens, while a further twenty-four people were injured by the devices.

Public ignorance about the bombs contributed to more deaths and injuries in the next few hours, even though loudspeaker vans toured the area at first light, warning against touching them. Keith Mann, aged twelve, was on his bicycle when he saw a butterfly lying on the ground. He threw it out in front of him and when it failed to explode he, as boys will, did so again. This time it did go off, but although he was only twenty feet away he somehow escaped serious injury.

A Mr Colbert was not so lucky. He had collected three bombs and had been told – erroneously – to make them safe by putting them in a water butt. When he found one, it was covered with wire netting. Needing both hands to remove this, he tried to give the butterfly that he had to a friend to hold, but the other man refused to take it. In the confusion the bomb fell to the ground and both men were killed. The other two bombs had been put down by a shelter door, where they later exploded and killed two more people.

Similar incidents occurred all over Grimsby. Two women were asleep when a butterfly came plummeting through the ceiling and dangled from the rafters of their room. One of them instinctively reached up to pull it down, and both were killed. At Canon Ainslie School a caretaker discovered a bomb and went to find some wardens. When he returned with three of them the device went off and all four men lost their lives. In all, sixty-six people were dead by the afternoon following the attack, and Kettles the undertakers had to send to nearby Brigg for more coffins.

With normal life suspended, No. 3 BD Company was sent into Grimsby and Cleethorpes to clear an estimated 1350 unexploded butterfly bombs. Its various sections had for once been gathered together, for assault course training at Nottingham, and so they were able to move quickly to the stricken areas. The company was commanded by William Parker, the officer who had seen the UXB at Lord's, and several days later Brigadier Bateman himself travelled north to monitor progress.

Two of Parker's junior officers were Eric Wakeling and Cliff Green. When they arrived at Grimsby they found the town at a standstill. The streets were pockmarked by craters and fires were still smoking. Hundreds of butterflies were hanging from fences and washing lines and garden gates. Some had penetrated slate roofs and were crouched on bedroom floors.

The awkward situations in which many of the bombs were lying made heavy demands of both the sappers' skills and their concentration. One of their first tasks in any area they surveyed was to try to pinpoint all twenty-three SD2s that would have fallen from their container when this was dropped. This was important since if one of the butterflies went off, there was a risk that sympathetic detonation could explode another that had not been spotted and was dangerously close to someone.

An early instance of this occurred when an officer blew up a butterfly that had buried itself in a road. The blast triggered another trapped in a roof gutter above him, which he had not seen, and he was injured. Elsewhere, Bateman found a butterfly trapped in a cupboard in a loft of the Grimsby Co-operative Society. A second was discovered under the roof. These were sandbagged and detonated. Three seconds later, a third bomb, about thirty feet away in another part of the loft, exploded but fortunately without causing serious harm.

Much ingenuity was deployed to remove or destroy butterflies that could not be handled or which were inaccessible. Green could remember reconnoitring one particular house: 'A bomb complete with wings was found lying on a double bed, with the clothes indicating the occupants had left hurriedly. A string was tied to the wings, the twine going out the window and down to the ground, in such a way as to drag it off the bed towards the window. A hefty pull was needed to take up the slack, but it went off very quickly.'

Complex machines made of pulleys and twine were rigged up to manoeuvre butterflies in to traps made of straw bales, but the soldiers' schemes did not always work. Wakeling, who was only twenty-one, had dealt with a group of five successfully, then 'I put a bit of string around the sixth one and pulled it. It didn't go off, so I pulled it again, and still it didn't go off.

'I was in a ditch with my driver and he said: "If you pull it much more, it will be in the ditch with us." So I went and looked at it. It was ticking, and I beat a hasty retreat. That was the stupidest mistake I've made. It would have gone off of its own accord if I had left it. It only had a few ounces of explosive, but it still killed you.'

Although the local police were a great help with reconnaissance, butterflies were adept at concealing themselves. They were found in kitchen gardens, sunk next to the potatoes, and Parker knew of several that had embedded themselves in damp lawns so that only the wings remained above ground. Only by a very slow and careful search could further tragedy be avoided.

'The worst one of the lot I found,' Green recalled, 'was a bomb cylinder on the top of the bedroom ceiling in an attic. It had left the wings in the roof slates and was lying between the ceiling joists. Although the wire had been pulled out, the fuze remained in the bomb.

'I had to kneel on the ceiling joists and noticed a space under one end of the bomb. I was sure that a clove hitch made by the boy scout double ring would be able to go over it. So I bent down again with the string in my hand, and as I got the clove hitch in position my knees slid on the edge of the rafter, and the bomb rolled forty-five degrees, then rolled back into position. This was about five inches from my face.

'I thought to myself, This is it. But nothing happened. I finished tying the knot, adding a half-hitch to make sure it was tight, took the string out through a ventilator and down to the ground. It only needed a small pull to set it off.'

Many of the butterflies had fallen in the countryside outside Grimsby, where the dangers were different. Given Britain's need to grow all the food it could, it was of some importance that fields, meadows and even undergrowth were cleared of the devices. Yet Nature did its best to hinder BD's efforts. 'Twelve of them, I remember, were in a pea field, marked by two-foot poles,' wrote Wakeling. 'But by the time we were able to return and deal with them, the peas had grown to two feet, six inches. We had a devil of a time trying to locate them.'

Climbing plants such as peas and beans presented particular hazards. Their heavy leaves, the tendency for tendrils to become entwined and the difficulty of parting these all increased the chances of exploding a bomb that had snagged unseen on some part of the crop. Searchers used two stout sticks to make a movement like the breast stroke of a swimmer, pushing forward and outwards a yard at a time. The arms were then swept backwards, and a complete scan made of the surrounding area.

Conversation was thought distracting and thus forbidden, but searchers were allowed a break every two hours, for tea and to rest their eyes. Many bombs were hard to see, since their vanes had been stripped off, and they were often half-buried and, using this method, a team of thirteen policemen and wardens took an entire day to sweep seven acres of dense clover. They found twenty-three unexploded butterflies.

'It took three months to clear them,' recalled Eric Wakeling. 'The city was almost completely paralysed. These small bombs were in attics, gutters, hedges, railway sleepers. People hardly dared walk in the street . . . The effect of these bombs was to bring the place to a halt. It was one of the best kept secrets of the war. The Germans never realised how effective they were.'

For, despite the casualties, the government gave very little publicity to the raids. Writing about the attack on Grimsby, officials made clear that 'now . . . we have to fear heavy concentrations of anti-personnel bombs', and it was thought preferable not to encourage these by revealing, through the use of warning posters, that butterflies were regarded as a serious menace. The tactic seems to have worked. Although they continued to be dropped sporadically along the East Coast, Grimsby remained the only urban centre targeted. The D-Day preparations, for instance, might have been severely hindered by the use of anti-personnel devices against the great numbers of troops massing under canvas on the South Coast. Yet only one small raid was mounted, on a camp in Dorset in May 1944.

There may have been another reason, too, for British reluctance to alert the Germans to the potency of butterflies. 'In view of our own plan to mix delayed-action anti-personnel bombs among incendiaries . . . the German attack on the Grimsby area on the night of 13th/14th June deserves special study,' runs one report on the raid of July 1943.

The Luftwaffe had sowed the wind, and now the RAF was ready to reap. Scientific analysis of bomb damage by Bernal and Zuckerman, among others, had confirmed for the head of Bomber Command, Sir Arthur Harris, a belief that he had formed during the Blitz that incendiaries were far more destructive than high explosive. Fire damage was more widespread, lasted longer and the heat cracked and distorted factory plant that resisted blast.

Plans were made to incorporate anti-handling devices into a proportion of the incendiaries being prepared for Operation Gomorrah, the heaviest series of raids yet launched against a German target. On 27 July, with much of its fire-fighting capacity disabled, Hamburg was engulfed by a firestorm, and burned as no city had ever burned before.

19

HITLER'S LAST HOPE

Forty-two thousand people died in the inferno that was Hamburg in that last week of July 1943. It had taken nine months in 1940–41 to kill as many in the Blitz. The combination of deadlier technology and bitter experience was rapidly perfecting bombing's potential for destruction. Roused to fury by the attacks on Germany, Hitler sought vengeance through the construction of still more fearful weaponry.

The Treaty of Versailles had banned research on heavy artillery, and one consequence of that had been to stimulate work by German scientists between the wars on rocketry. As early as 1939, Hitler had vowed to build 'a weapon with which we ourselves could not be attacked', and at Peenemünde on Germany's Baltic Coast the Luftwaffe began to develop the next step in its arsenal – bombs that could pilot themselves.

The first *Vergeltungswaffe* or 'vengeance weapon', the V-1, was designed by the Fieseler aircraft company. It was shaped like a jet, twenty-five feet long with a short wingspan of sixteen feet, and powered by a rear-mounted pulse engine whose distinctive insect-like hum would lead the British to dub it the 'buzz bomb' or 'doodlebug'.

The rocket was held on course by a gyrocompass while a simple odometer linked to a propeller driven by airflow determined the distance the missile had flown. Once a preset number of revolutions of the propeller had been counted, the V-1 was tilted into its final dive. This change in elevation cut off the fuel supply. The sudden lack of noise from the engine brought fear to those on the ground, as they waited to see where the 'flying bomb' would fall.

By early 1943 the existence of the rocket was suspected by British experts, among whom was the thirty-one-year-old R. V. Jones, a physicist who had played a leading part in disrupting the beams used by the Luftwaffe's bombers to find their targets. A raid in November by the RAF on the storage complexes in northern France being constructed for the missiles slowed development of the Nazis' main hope of turning the tide of the war. Not until 12 June 1944 – nearly a week after D-Day – were the first V-1s ready for launch. From more than a hundred well-camouflaged sites in the woods of Normandy and from bunkers in the Pas-de-Calais, the Germans prepared to dispatch their wrath across the Channel at 340mph.

The V-1 was intended to inspire terror. It carried a warhead of 850 kilograms of amatol, a mixture of TNT and ammonium nitrate, but it was the alien nature of its delivery that made it seem more frightening than a conventional bomb. People had become accustomed to the Heinkels and Dorniers, which at least were piloted by other human beings. While the V-1 was no more random in its selection of victims, its sinister shape silhouetted against the sky and the agony of waiting to see where it was going to land, made it seem more perverse and inhuman. And unlike the bomber crews, it did not need to take heed of enemy fighters and radar, so could strike at any hour.

Most of the missiles aimed at Britain fell on London, which had grown used to feeling safe from bombing in daylight hours, and it was perhaps the resumption of this that led the V-1 to be regarded with particular fear, and tested the nerves of war-weary

Londoners more severely even than the Blitz. Fearful of the effect that it might have on the population, the government initially kept secret the onset of the new weapon.

The first V-1 to reach London fell on 12 June on a railway bridge at Mile End, smashing the tracks and killing six civilians. No publicity was given to this, and the BBC was not allowed to report the existence of the rocket until Friday 16th.

When Elisabeth Sheppard-Jones, a young Welsh woman, attended a service at the Guards' Chapel that Sunday, few in the congregation were yet attuned to the danger: 'In the distance hummed faintly the engine of a flying bomb,' she wrote later. '"We praise thee, O God: we acknowledge Thee to be the Lord," we, the congregation, sang. The dull burr became a roar, through which our voices could now only faintly be heard. "All the earth doth worship Thee: the Father everlasting."

'The roar stopped abruptly as the engine cut out. We were none of us then as familiar as later all London and the south was to become with Hitler's new weapon, to recognise this ominous sign. The Te Deum soared again into the silence. "To Thee all Angels cry aloud: the Heavens, and all the Powers therein." Then there was a noise so loud it was as if all the waters and the winds in the world had come together in mighty conflict, and the Guards' Chapel collapsed upon us in a bellow of bricks and mortar.' She was trapped in the debris for several hours before being freed, but she had been severely injured and lost the use of her legs. One hundred and twenty-one civilians and soldiers were killed.

R. V. Jones was working near by, and went down to the chapel to see the damage for himself. 'I knew its warhead was going to be about a ton. I knew what a ton of explosive could do . . . But it struck me then how very different the academic appreciation of explosions was from the actuality . . . One lasting impression I had was the whole of Birdcage Walk was a sea of fresh pine tree leaves, the trees had all been stripped and you could hardly see a speck of asphalt for hundreds of yards.'

Within three days of the first V-1 landing on London almost five hundred people had been killed by 647 of the bombs. The spluttering noise of their engines, and the fiery trail they left at night, soon became familiar to those in the capital. As fireman Cyril Dermarne recalled, so did the devastation signalled by the rise of a sooty plume of smoke into the leaden June sky.

'A particularly nasty, gory situation confronted us following a V-1 explosion in Dames Road, Forest Gate. A trolley bus, crammed with home-going workers had caught the full blast and the whole area was a sickening sight. Dismembered bodies littered the roadway; others were spattered over the brickwork of the houses across the way and the wreckage of the trolley bus was simply too ghastly to describe . . .

'The roof and upper deck, together with the passengers, was blasted away. Standing passengers on the lower deck also were flung against the fronts of houses on the other side of the road. The lower deck seated passengers were all dead.' Although many of the victims had been decapitated, they were still sitting down, as if waiting to have their fares collected.

Although the V-1 attacks had long been expected by the government, it knew few details about the bomb's workings. From BD's point of view this was no handicap as the missiles at first added little to their task since they exploded on impact. Then, eleven days into the V-1's campaign, one landed largely intact at Strawberry Hill Farm, near Staplecross in Sussex. Here was a chance to uncover some of its mysteries, and on 24 June John Hudson was sent to take a look.

From parts of other rockets that had previously been recovered, Hudson knew the general configuration of the missile. The charge sat forward of the wings, and in the nose was an electrical impact fuze, the Type 106X. A second kind of fuze, the (80A), had also been found. This was a mechanical device intended as a back-up, which would fire whatever the angle at which the weapon hit the ground.

The V-1 at Strawberry Hill Farm was lying in a wood with the

(80A) visible in a side fuze pocket. As Hudson immediately saw, behind it was a second pocket containing a third fuze whose presence had not been suspected and whose function was unknown. Having reported this to Bateman, he was ordered to retrieve it without fail: 'every possible step must be taken to ascertain the fuzing system, however long it may take to do so'.

Hudson now had three tasks: he had to establish the nature of the new fuze; to devise a method of safely obtaining it; and to accomplish both of these aims without exploding the warhead of the V-1, losing the remains of the missile and the fuzes, and perhaps his own life. The first step was to identify how the fuze worked, and for help with that Hudson turned to John Dawson and his 'field photography' unit.

'It must have occurred to someone,' said Hudson later, 'that it would be useful to X-ray bombs. But it is not easy to do so through a ton of metal as you get terrible dispersion of the rays. And it had to be something that was compact to use, not an X-ray machine . . .

'A way was found: we were given a tiny source of radium. It was kept in the caves at Chislehurst . . . There was highly radioactive gas coming off it into a little pipe at the top. If you put a glass tube filled with mercury − the size of a match − on it, the gas came bubbling out, this pushed the mercury out, and when the mercury had all been pushed out you just broke it off and sealed it with a Bunsen burner. It was filled with radon gas. It decays very rapidly, it has a half-life of four days or so . . .

'This thing was broken off and put into a fist-sized lump of lead, with a strong brass screw to close the hole. You could then carry it. If you want to take a picture through a bomb, you go to a local hospital and get an X-ray sheet. You put it on one side of the bomb, prop the lead up against the other side, undo the screw and leave it.'

Over the next few days, improvising as best he could in the conditions, Dawson established that two of the fuzes were those encountered before, and that the third had a clockwork movement similar to a (17). The X-rays also showed that there was no anti-withdrawal device behind it. The question was now how best to defuze it, given that it was thought to be armed and had a short time setting. Hudson proposed to jam it with a Stevens Stopper, but wanted to be certain that there was a channel within the fuze that would allow the viscous solution to surround the cogs and wheels. Some way had to be found of getting the film and the radium seedling closer to the fuze in order to obtain a clearer picture. That would inevitably mean cutting in to the thin steel of the bomb. Any vibration might start the fuze ticking, which ruled out the use of a drill or a saw: what else could be used to slice metal?

With Hudson at the site was Bob Hurst. 'Bob and I decided that rather than use a chisel – so as not to cause any vibration – if we dribbled acid on a semi-circle of the case it would start to eat through it. When it had done that, we could turn back the casing and start to wash out the explosive.' Suffolk and Hudson's work with acid – abandoned several years earlier because of its dangers – was now to come into its own. The only difficulty would be in trying to find laboratory-strength acid in the middle of rural Sussex. Then inspiration struck.

'We went to a girls' school in Hastings and asked if we could see the headmistress. She came looking very sour, thinking one of our soldiers had got a girl into trouble. We explained that we were very secret, couldn't tell her why we wanted it, but we did want a bottle of nitric acid if she could let us have one. She looked very relieved and called the science mistress in, and she found a bottle for us.'

The long summer evenings allowed work to continue for much of the day, interrupted only by other V-1s falling near by, for the wood lay beneath their main flightpath to London. Late in the afternoon of 29 June Hudson and Hurst started to use a thin jet of acid to make a long v-shaped cut around the fuze pocket. Owing to their past experience at Richmond Park, they were acutely aware of the dangers of any acid coming into contact with almost a ton of high explosive. They therefore proposed to stop the process before the cut was through and to complete it with a sharp steel blade.

By lunchtime the following day, despite the distraction of two rockets being shot down over them by fighters, the pair had scored the metal for more than a foot, allowing them to fold back a flap of steel and expose the main charge. With warm acetone and hot water they then began to dissolve the white biscuits of explosive around the fuze pocket to make room for the radium.

The work continued the following morning, 1 July, but overnight both Hurst and Hudson had fallen ill. Hours of contact with the chemicals and the toxic fumes emitting from the bomb had caused di-nitro benzene poisoning, leaving them with blue lips and a leaden hue to the skin, as well as headaches, extreme tiredness and nausea. Hudson felt so sick that he could only carry on for a few minutes the next day, and Dawson then had to relieve Hurst when he too became incapacitated. Both men drank copious draughts of milk to try to alleviate the symptoms, but did not fully recover for the best part of a week.

Meanwhile, Dawson had managed to set up the photographic plate and the radioactive source either side of the fuze pocket. This allowed him to obtain very clear radiographs which confirmed that the clockwork mechanism was identical to a (17B) long-delay fuze, giving it a maximum running time of two hours. It was impossible, however, to gauge what the setting was on this particular fuze, or how long might be left.

The next step was to extract it. It had been decided that it would be safest to attempt to remove it gently by remote control, without first trying to immunise it, in case the vacuum pressure detonated the fuze. Similarly, any strong movement while it was being retrieved risked freeing the clock, which might only have a few seconds to run. All their work of the previous week might be destroyed in an instant.

By 2 July Hudson and Hurst were well enough to begin construction of the most sophisticated and ingenious of all fuze extractors. First, a fuze key was fitted to a dummy head screwed tightly on to the fuze itself. Hudson, operating steel cables, worked the key from a hundred yards away. The key was also suspended from overhead lines and cushioned by a cloth, so that when the fuze came out the key was swung clear without hitting the bomb or the ground.

As the fuze appeared, it was then drawn into contact with a hollow magnet by a string from the key. This would immediately be switched on so as to prevent the clock from running should it have been re-started by the sudden jerk. Hurst would be listening on the stethoscope in a trench fifty yards from the bomb to confirm all was well.

They held their breath, and in the evening light Hudson gave a tug on the cable. It worked perfectly. He was able to saunter over to the fuze, now held in the magnetic field, and remove the gaine. When the fuze was later tested it was found to be fully armed but not started, and set to function in thirty-two minutes.

Hudson had felt under great strain, but he was right to have been so cautious. In October a fuze pocket that had been picked up from the remains of a V–1 exploded without warning, killing a BD officer and wounding two others. The most probable explanation for the disaster was that the simple act of moving it had re-started the fuze's clockwork mechanism.

By then Londoners' nerves were being tormented by a new terror weapon. The V–2 began to fall on the capital in September 1944. Launched from mobile sites in France and Holland, the rockets were forty-six feet long, carried a warhead of 994 kilograms and travelled at the then-incredible speed of 3600mph, more than four times the speed of sound. From take-off in Europe to impact in London took less than three minutes, and as they were faster than sound there was no warning of their arrival. The explosion produced a shattering bang that could be heard ten miles away, and a single V–2 could level an entire street of houses. Four that landed on Croydon damaged between them two thousand homes.

The destruction that they caused brought renewed fear to parts of South and East London that had already suffered greatly from bombing, such as Lewisham, Woolwich and Hackney. Their sheer unpredictability had the potential to generate panic among Londoners, who would have no warning or chance to see shelter, and fearing their effect on morale the government imposed a blackout on reports of the damage that they did. Rumours inevitably circulated, but the restrictions on the press were not lifted until November.

More than 1400 V–2s were to strike Britain, all but a handful landing on London. Yet only four failed to explode, and all of those were in the last weeks of the attacks. They were all dealt with easily enough by BD, but by then 8958 people had been killed by V-weapons, and almost another twenty-five thousand had been injured. In the autumn of 1944 news came through of recognition for BD's work at Staplecross. Dawson and Hurst, both members of Gough's team, had been deservedly awarded the George Medal, while Hudson had got a second. He was one of only three RE Bomb Disposal officers to win the GM and Bar during the war.

A few months earlier, he had written to Lord Rothschild, the scientist and MI5 officer, to congratulate him on his own GM. Victor Rothschild was in charge of counter-sabotage for the Security Service and, having occasional recourse to defuze bombs himself, was in touch with Hudson about technical developments, though fully aware of the gulf in expertise between them.

'I feel very sincerely,' he wrote to Hudson, 'that I have some form of apology to make to you and the other professionals for my award. Having seen tremblers, mercury switches and other anti-withdrawal handling devices, I am fully aware that the few bombs I have dismantled in this war in no way compare with the jobs that you and your people do.'

It was a sincere tribute from one brave man to another, but outside BD very few knew of the exact risks that were taken by its personnel every day. John could not even share his worries with his wife, despite the fact that while working on the V-1 he had been able to return home every night to nearby Burgess Hill, where she and the children were living. He would arrive covered in mud, and she could only wonder at what he had been up to. The constant strain took a toll on both, though without affecting the deep bond that bound them.

'Oh John,' wrote Gretta after one brief weekend they had spent together, 'I feel so overwhelmingly John-sick, those precious few days together left a deeper mark than I'd imagined – your devoted love fills me with humility and painful longing . . . I only hope the days will fly by and that my heart will ache a little less each day.'

20

SING AS WE GO

It should have been a simple job. Ken Revis had, in three years, been transformed from a nervous recruit defuzing his first bomb with a piece of cord and a folded newspaper into an expert BD officer. Now, in September 1943, with the threat of invasion defeated, he had been assigned to make safe the booby traps planted earlier on two symbols of a more peaceful Britain, the West and Palace Piers at Brighton.

Since 1940 no one had trodden the deck of the Palace Pier, which had been isolated from the beach by a hundred-foot gap. Like a great rock of steel and wood, it jutted out of the sea, connected to the mainland only by an undersea cable that linked a trigger in the town's aquarium with a mass of depth charges placed beneath the concert hall on the pier. Revis made a careful reconnaissance of its struts from a rowing boat, then set about dismantling the explosives. Once done, he turned his attention to the long West Pier.

It was a sunny morning, and it was relatively easy to spot the small blue dots on the planking that marked the location of the anti-tank mines hidden underneath them. 'The procedure,' recalled Revis, 'was to locate the mine, and saw a small hole in the

planks. Then you would reach down with your hand and block
the detonating mechanism' – a spring-loaded striker – 'which was
an extremely sensitive piece of equipment.'

The mines had been laid in case German troops used the pier
as a landing ground, and Revis and his NCO, Corporal Marnock,
quickly dealt with six of them. Stooping a little, Revis looked for
the seventh dot, which was about the size of a match head. 'I
straightened slightly to stretch my back. I said to the corporal who
was behind me, "This is money for old rope." I lowered my body.
I was virtually kneeling on the mine. It blew up.

'I know I wasn't moving at the exact moment of the explosion.
The mine, and twelve others, had gone up simultaneously.'
Marnock was only about four feet away, but Revis took the full
force of the blast. The timbers caught fire, part of the decking collapsed and in the thick black smoke the shocked Marnock could
not find his officer. For several minutes Revis lay there alone, with
appalling injuries to his arms, chest and face.

'I don't remember being unconscious. I felt I had been cuffed
on both ears by a heavyweight boxer.' Eventually he was lowered
by a bosun's chair into the boat and ferried to shore. 'I remember
being on a stretcher and hearing an Australian nurse say, "Cover his
face." And I said, "Take that bloody thing off, I'm not dead yet!"'

Revis was taken to the Queen Victoria Hospital at East
Grinstead, which the plastic surgeon Archibald McIndoe had
turned into a pioneering centre for the treatment of burns,
notably those suffered by pilots in the Battle of Britain. Revis was
to endure twenty separate grafts and operations while his face was
reconstructed, with bone being taken from his rib to rebuild his
nose. Nothing, however, could restore his sight. His war was over,
but a life-long struggle was just beginning.

Although bombs did not always kill, they frequently exacted
a toll in other ways. Towards the end of 1944 John Hudson's
health began to break down. He developed enteritis, then had a
severe attack of bronchitis and was confined to bed for a month.
The underlying cause of these ailments was poisoning due to

inhalation and contact with chemicals over many months, but he believed that the cumulative stress of the work had also caught up with him.

'I think I was run down because of the nervous strain,' he admitted. 'It was a pretty tense sort of field I was working in. I don't think I was getting scared, but I'd worked pretty hard. I had been in since the beginning. We worked all the hours that came, if there was something on.

'We were operating from an office, but Bob Hurst and I, and the chaps from the Navy and RAF, if there was anything to do we didn't think of knocking off at five-thirty . . . I did get leave during the war. I used to go home Friday night to Sunday night about every three weeks . . . I was very lucky compared with chaps who went to Egypt and never came home for years.'

By the start of 1945 they were starting to come home, though. There was a final spasm of attacks against London, particularly by V-2s, in early March, but the last bomb to be dropped on Britain landed on Sittingbourne at the end of that month. Four weeks later, Hitler was dead.

The Civil Defence organisation was stood down in early May. BD had already contracted its operations, notably in London. Well before the Allies entered Berlin, London was protected by just ten sections, the same as had been stationed there prior to the Blitz. Many squads had been sent overseas or given the hazardous and painstaking job of clearing the country's coast and beaches of the 350,000 mines laid in 1940. Another 72,500 had been put down by the Germans in the Channel Islands.

The effects of wind and tide meant that maps of the minefields were no guide to where the devices might now be found. Sand and pebbles had to be probed slowly and cautiously, as many of the mines required little pressure to set them off. By March 1946 280,000 had been removed, sometimes with the aid of German POWs, but at very heavy cost. More than 150 BD personnel were killed by mines between 1940 and 1948, and another fifty-five injured. The clear-up would continue until 1972.

About 147,000 British civilians, including civil defence workers, were killed or injured by bombing during the war, the great majority during the Blitz and the V-weapon attacks. The dead numbered some 60,595. (For the sake of comparison, fifty-five thousand aircrew lost their lives serving with Bomber Command, which had the highest rate of attrition of any British unit.) For more than two thousand days, the United Kingdom had been under aerial siege, with London receiving the bulk of an estimated 68,500 tons of explosive delivered by the Luftwaffe or by rocket. Again, for comparison, the Allied air fleets dropped almost 1.2 million tons on Germany in 1944 alone.

Having begun from a standing start, by the end of the war BD had defuzed about forty thousand High Explosive (HE) bombs in Britain, together with some 5700 butterfly bombs and another 6900 anti-aircraft shells, incendiaries and other types of ammunition. Tens of thousands more weapons, Allied and Axis alike, had been dealt with on battlefronts and in munitions dumps in the Far East, North Africa, the Mediterranean and during the advance through North-West Europe. This work would also continue after the war. Initially it was thought that all bombs in Britain would be found by the end of 1946, but between 1950 and 1958 140 HE UXBs were disposed of, and 150 new ones reported.

There is no consensus as to the exact total of casualties suffered during bomb disposal operations. Different records cite different incidents. Some include civil personnel such as scientists, policemen and factory workers who were involved, while others omit those killed clearing British mines. A conservative estimate of all those who died while helping to dispose of wartime bombs in the UK and abroad would be about 750 people, of whom at least six hundred were from the Army. Many others were maimed or grievously wounded.

While the war in Europe was over, there was still plenty to occupy the minds of BD's experts in its aftermath. In particular, Herbert Gough's team were interested in what could be learned from their German counterparts. Bob Hurst made requests to BIOS – the British Intelligence Objectives Sub-Committee – for two particular groups to be rounded up and interrogated: the fuze engineers of Rheinmetall-Borsig; and the leading German bomb disposal officers, or *Feuerwerker*.

Among the latter was Major Eric Renner, an adjutant in the Luftwaffe who had been captured by the Americans in Normandy. He had married a French girl and was at pains to stress that he had never been 'in accord with the ideals of Nazism'. He was only too happy to reveal the German approach to bomb disposal.

The basic unit consisted of an officer and three or four NCOs, all specialist armourers who had been through a lengthy training course. Before the war more than a thousand of these were sent each year to the ordnance school at Halle, but only about half passed the exam.

The principal difference in method between the British and the Germans was that the latter used forced labour to do their digging. Renner had looked after five airfields in France, and had had three hundred French civilians working for him. Later, when the bombing of the Reich intensified, Himmler had authorised the SS to press large numbers of convicted criminals and inmates of concentration camps into service. Some were promised early release if they volunteered for such squads.

At first there were examples of this agreement being kept, but an end was soon put to this. Word got around of the dangers, and though rumours persisted of privileges for such prisoners, by 1943, for instance, only six out of fifty inmates assigned to a disposal team at Kalkum were volunteers. Russians were singled out for the most dangerous tasks, and the casualty rate was high, not least because the Germans were careless with prisoners' lives. At Bochum in August 1943 a dozen of them were ordered to tow a car chained to a bomb thought to have an anti-motion fuze. All were 'blown sky-high'. One inmate, Jan Jakubowski, had gone to fetch water when another bomb exploded, killing thirteen of his companions: 'I brought them all back in a couple of buckets . . . they were torn to pieces.'

There was much work for them to do. Between July 1943 and March 1944 SS Construction Brigade II, formed of camp prisoners, cleared more than 7850 UXBs – a quarter of the total dropped on Britain in the war – most of them from the devastating raids on Hamburg. Renner calculated on '10 per cent of all bombs dropped being unexploded, but that the percentage of American duds was considerably higher than that of the British.

'He cited one instance occurring at Le Bourget field in June or July 1943 when there were ninety unexploded American bombs out of 450 to 500 dropped . . . He stated, and seemed quite incensed about it, that nearly all of the failure of American bombs was the fault of either the fuzing or dropping. It seemed to him like gross carelessness to drop bombs with the arming wires in them, or bombs without any fuzes.'

As in Britain, it was officers who took the responsibility of disarming the UXB, but they had no centralised hierarchy to help them. There was no equivalent of the UXBC, and individual officers largely developed their own equipment and relied on their own experience. Their numbers were slightly higher than across the Channel, with about six hundred officers active in Germany. Curiously, their chief complaint was that their own air force was reluctant to share information on its own bombs and fuzes.

In the early years of the war they had been able to leave the relatively small number of Allied UXBs for ten days before blowing up in situ those which remained. As Goering's boast that none would land on the Ruhr was made to sound more and more hollow, there was a switch in strategy, with the emphasis now on defuzing, and therefore more risks being taken – the reverse, of course, of the pattern of BD's war in Britain.

Many of the techniques and apparatus evolved to deal with the later flood of UXBs were developed by the best-known of the *Feuerwerker*, Heinz Schweizer. The first officer from a

Spreng-Kommando (bomb disposal squad) to receive the Knight's Cross, Schweizer had led the team that tackled the giant UXBs left behind by the Dambusters raid in May 1943.

The foremost name in Berlin, where the valiant and desperate efforts of *Feuerwerker* were eventually overwhelmed by the sheer volume of bombs falling, was Egon Agtha. In his early twenties, Agtha had been severely injured when a UXB exploded in 1941, but having been discharged from the military he returned as a civilian volunteer. In the space of three months in early 1945, Agtha won three of the highest German decorations for bravery, but he was shot on 2 May while trying to escape from the Soviets across the Charlottenbrücke.

Renner knew little about the British BD set-up or their methods, but in any case considered the (50) and the Zus 40 impossible to disarm. The Rheinmetall fuze engineers knew otherwise, and in May 1945 Merriman was among the first to enter the firm's works at Unterlüss, a key test centre for proving guns and ammunition. Documents found there revealed much of the structure of the company, and an alert was put out to apprehend those who had created the electric fuzes. Many of Rühlemann's research staff were subsequently captured at Mulhausen and interrogated, not so much to learn about what they had done as about what they were currently experimenting on. For the game had changed, and with it BD's focus. What was needed now were advantages in ordnance research that could be used against a new enemy: in September 1945 Hurst found a new file on his desk; in it were details of Soviet bombs and fuzes. The Cold War had begun.

Bob Hurst would go on to become the first director of the Dounreay nuclear power station. For many, the war acted as a catalyst that would take them in completely unexpected directions. Others quietly returned to their previous occupations, albeit to a different world. All were marked deeply by their experiences in BD.

Stuart Archer was not demobbed until January 1946. He had moved on from Thame in 1943, to a staff job at Southern Command based at Wilton House, outside Salisbury. Then, as a major, he had commanded 12 BD Company at Horsham, clearing the South Coast of the remainder of its mines. It was a job well suited to his conscientious nature: 'When you're coming back to peacetime, you can't leave a minefield for little Johnnie Clark to tread on while on holiday.' There were no fatalities in his company.

He resumed the reins of his architectural practice, which had been kept going by his partner throughout the war. A few years later, however, he joined the Army Emergency Reserve, and from 1963 until 1967 was Honorary Colonel of the Bomb Disposal regiments. He still lives in London.

Jack Howard was buried in the churchyard of St John the Baptist in Charlton. Stories about his life became legend, most notably in Michael Ondaatje's novel *The English Patient* which, while a fine work of fiction, gives a portrait of Suffolk – the mentor of the bomb-disposing Sikh protagonist – that bears little relation to life. Suffolk's BD work is commemorated in a stained glass window in the church of St John, above which runs a verse composed about him by John Masefield, then Poet Laureate:

> He loved the bright ship with the lifting wing,
> He felt the anguish in the hunted thing,
> He dared the dangers that beset the guides
> Who lead men to the knowledge nature hides.
> Probing and playing with the lightning thus,
> He and his faithful friends met death for us.
> The beauty of a splendid man abides.

His eldest son is the present Earl of Suffolk.

Given his ill health, John Hudson was able to apply for release soon after VE Day. He had been offered a position with the New Zealand Department of Agriculture, and after receiving the second of his George Medals from the King, the Hudsons made the month-long voyage to Wellington. The outdoor life suited John and his health soon returned to its customary robust state. Other strains took a little longer to overcome: 'My wife told me often that I was a bit difficult to handle for the first year or two after the war. And I got to know my two little boys, because I hadn't seen much of them.'

The war had altered Hudson's outlook and prospects. It had given him more confidence, made him a first-class communicator – there was no room for ambiguity in bomb disposal – and accustomed him to mixing with those who before 1939 he would have regarded as his social betters. Despite his modest background and degree in horticulture, he had been taken seriously by senior officers and distinguished scientists, and had known as much as them about how bombs worked.

He may perhaps have regretted not going on to study physics, but he was to achieve much in his chosen field. In the fifties he became the first Professor of Horticulture at the University of Khartoum, and from the sixties was the eminent director of the Long Ashton research station and Professor of Horticultural Science at Bristol University. Though proud of his war record, he never wanted it to overshadow what else he had done, and it was a side of him to which his students and neighbours remained oblivious. He died in 2007, aged ninety-seven. Hudson House, the headquarters of 101 Engineer Regiment (Explosive Ordnance Disposal), the RE's only TA BD unit, is named for him.

In May 1945 the Minister of Supply, Sir Andrew Duncan, received a letter from the chairman of Unilever, Geoffrey Heyworth. The industrial colossus was searching for 'an engineer of the highest calibre to have general supervision of the engineering aspects of all our processes . . . throughout the world'. Its eye had alighted on Herbert Gough.

Duncan commended him as 'one who is likely to serve you faithfully and well', and it was characteristic of Gough that his valedictory letter should express his concern for those who had done likewise for him. 'As a player of games, I have always endeavoured to have relations with my immediate colleagues which were of the "team" and "family" type', he wrote. 'I hope that you will be able to arrange for official recognition of the part they have played and the splendid work they have done . . . I feel that, in comparison with other Ministries and Departments, the MOS has not yet recognised sufficiently the work of its scientific and technical staff.'

Their fate, as had been his, was to be overlooked. The most celebrated scientific names of wartime were not those who worked in its engine rooms, running the research divisions, but the advisers who had the ears of ministers. The likes of Lindemann and Blackett and Bernal had prospered, while Gough's role had been steadily downgraded after more reshuffles at the Ministry, so that by the end of the war many of his executive responsibilities had been distributed elsewhere.

Of course, in the first years of the war the DSR had been in a privileged position. Research and development had been vital to Britain's ability to hold its own. After Alamein, and the arrival of the Americans, that object had been achieved and it was relegated to its usual place, subordinate to the demands of production. In truth, by 1945, Gough was tired of fighting on more than one front and not averse to a change of scene.

The Ministry paid tribute on his departure to his key role in the creation of the Scientific Advisory Council. This had 'made available to us the advice and service of many of the most eminent scientists and engineers in the country, and its stimulating influence has been felt in every department'. The official history of the MOS was more generous yet in its verdict, saluting the professional approach that Gough had brought to the hitherto amateurish and conservative world of military science: 'The Director General of Scientific Research & Development of the period defined the duty of research and development as that of ensuring that every item of military equipment for which it was responsible did actually represent the latest and most complete embodiment of all that science and technology, design and the modern production method could individually and collectively offer . . .

'When one compares the freedom and authority enjoyed by civilian scientists and technicians, both at headquarters and in the establishments, in the period 1943–45, with the isolation and subjection of pre-war days, the change is among the most striking features of the organisation of British war production . . .'

Special mention was made, as Gough left, of his 'enthusiasm and undaunted cheerfulness', which had helped to make such a success of his direction of the scientific aspects of Bomb Disposal. He chaired the final meeting of the Unexploded Bomb Committee in October 1945. The intellect, perspicacity and, on occasion, personal bravery of its members had also contributed greatly to the blunting of a dagger that had been aimed at the main artery of the war effort, Britain's industry and infrastructure.

Herbert Gough was engineer-in-chief to Unilever for a decade after the war, and in 1949 was elected President of the Institute of Mechanical Engineers. In June 1965 he had just finished a round of golf when he collapsed and died almost instantly. He was seventy-five.

Herbert Rühlemann's later career had taken a rather different course. At its zenith, his importance to the Reich had been such that it had been able to overlook some awkwardness in his private life. For some time, he and his wife Sophie had been growing apart. Her own career as a writer had stalled after her refusal to allow the Nazi Party to republish some of her books, whose mixture of folklore and family values appealed to the leadership. As a result, her books had been banned and burned.

By the mid-thirties, Rühlemann had already met the woman who would become his second wife. Heidi bore him a daughter in 1941, although he was still married to Sophie. His position at Breslau was nonetheless untouchable, not even by this scandal. For several more years he continued to work on new projects, including fuzes for rockets. He also refined those he had already devised.

It was not so much intelligence about British BD that drove this as the imperatives of making fuzes cheaper and simpler to produce.

'A typical case', he wrote in his autobiography, 'was the redesign of the impact switch in the [electric] fuze. In its original design this switch consisted of two brass balls about one eighth of an inch in diameter soldered to a phosphorous-bronze wire . . . The improved design replaced both balls by a tight cylindrical spring with an increasing spring diameter. A one-piece design, no assembly of balls to the deflection wire, no soldering and the spring-forming machine produced such a spring at a rate of several pieces per second. There was no plating required, a plated wire was used. And there was no scrap.'

He continued to travel to evaluate the performance of fuzes in different soils, but insisted later that these occasional trips to Occupied countries 'really had nothing to do directly with the war'. Not until he visited Italy, where he noticed that he had no need of Italian because everywhere he stopped there were German soldiers, did he begin to think that the war might not be won: the Italians had signed an armistice with the Allies.

And Rühlemann, too, was feeling undervalued. In late 1944 the German government informed Rheinmetall that it was going to stop paying out on Rühlemann's patent for the electric fuze. For the first time, he learned that he was entitled to 3 per cent of the fees, and so far his employer had had twenty-eight million marks – and with overseas sales, he calculated, as much as forty-three million (one billion pounds today). Rühlemann reckoned that he should have been paid 1.25 million marks, but so far had only received 180,000.

He soon had other things to worry about. In May 1945, with the Russians advancing on the plant on Breslau, he was forced to abandon everything and flee westwards, where he was picked up by an American intelligence unit. He was held first near Versailles, together with leading figures in the regime such as Albert Speer, and then at Kranzberg Castle, the interrogation centre known by the Allies as 'Dustbin'. By July 1945 Rühlemann was in London with two other key figures in bomb technology, his chief Luftwaffe contact General Marquard and Wernher von Braun, the architect of the V-2.

The British were keen to make use of Rühlemann's knowledge and, although he spent a year in prison and then for several years had to take low-paid engineering jobs in Germany to make ends meet, by the late forties he found himself back in demand. He was headhunted by the US War Department, and in 1948 quietly given a job at the Naval Ordnance Laboratory in Maryland. Like dozens of other German scientists who had been instrumental to the Reich's success in battle, he was now to be employed by the Americans in the new fight against their common foe, Soviet Russia.

He and Sophie had finally divorced, and in 1949 Heidi and their daughter joined him in America. He was happy in Maryland, where essentially he was continuing the work he had done for Rheinmetall, and was surrounded by other Germans. He did not forsake his European habits. On Sundays, he wrote his letters, and even when cleaning the yard wore a jacket and tie. Once his contract came to an end in the mid-fifties, one of the great military engineering geniuses of the twentieth century finished his working life at a small company in Philadelphia that made automated equipment. Dogged by heart problems, Rühlemann retired in 1968. He spent much of the next eighteen years writing his memoirs, *Father Tells Daughter*, trying to justify aspects of his life that at times he seems to have found hard to rationalise to himself. He died in Pennsylvania in 1986.

After returning from the Middle East in 1944, Arthur Merriman, GC, had continued his connection with BD, working at the armaments research centre at Fort Halstead for several years. Soon after the war's end, he had written a report for the War Office on the so-called V-3. These were giant batteries of superguns sunk into underground sites in the Pas-de-Calais and trained on London. Each gun barrel was 425 feet long and by means of multiple charges fired along the length of this as the projectile passed, the shells were to be thrown as far as one hundred miles. The scheme was thwarted first by technical difficulties and then, in July 1944, by an RAF raid that damaged the V-3 complex at Mimoyecques, near Cap Gris Nez. Merriman took part in the site's demolition at the end of the war. In 1948, he became Registrar of the Institute of Metallurgists. He died in 1972.

In early 1943, as the need for a large BD organisation began to fade, conscientious objectors such as Victor Newcomb were offered the chance to transfer to the Royal Army Medical Corps. Like many other conchies, he found the opportunity to train as a stretcher-bearer with the Parachute Field Ambulance compatible with his humanitarian aspirations, and preferable to a reversion to labouring duties with the Non-Combatant Corps.

On 5 June 1944 he was dropped into Normandy a few hours before D-Day. He was later taken prisoner and spent two months at Rennes working in a German-run hospital staffed by British medical POWs. After he was liberated by Allied troops, he took part in the final assault across the Rhine. Newcomb married a French nurse whom he had met in Rennes, and in civilian life became a teacher.

Not long after the disaster at Castle Street in Swansea, James Lacey's section was visited by a major-general. Lacey was the only NCO on hand to show him around, and so impressed was the general by his briefing that he told him to apply immediately for a commission on his recommendation. Lacey was posted away from BD for officer training, and ended the war in Burma as a staff captain in the Transport Directorate of the 14th Army. He died in 2008.

Bert Woolhouse was discharged by the Army after the explosion that killed Sapper Ash and his other comrades. Looking back years later, he still felt that he had missed out: 'I wanted to try and get on in the Service. As a matter of fact, I'd have loved to have been a regular soldier . . . I still envy a regular soldier, I mean that.

It's a wonderful life.' He and Jane settled again in London. He died in 2009.

Following the accident at Brighton, Ken Revis never recovered his sight. Gradually he learned to cope with his disability, mastering Braille and then touch typing. He was sent to India to represent the St Dunstan's charity for the blind, and later became a press officer with Morris Motors. He went gliding, water-skied and drove a sports car at 100mph at an airfield, with his wife Jo giving him instructions from the passenger seat. Finally he qualified as a solicitor. Revis considered himself 'just an ordinary person who cannot see', and thought himself fortunate at least to have had vision before he was injured. Among the many committees of which he was a member was one which lobbied for the restoration of Brighton's West Pier. He died in 2002, aged eighty-four.

On being demobbed, Len Jeacocke was offered three times his salary to put his skills to good use, disposing of mines in the Egyptian desert to enable a company to prospect for oil there. 'I dithered and put off a decision,' he confessed. 'Eventually the chance slipped away. I lacked the courage I suppose.' He went back to his former job and later became a supplies officer for Eastern Gas. Before his death in 1999 he was also Secretary of the Royal Engineers Bomb Disposal Association.

By the middle of the war, Wally Fielding was getting pains in his chest. He had an interview with someone he believed to be the regimental medical officer – 'I didn't know anything about psychology in them days.' He was sent to London and examined by Hans Eysenck, the German-born psychologist and expert on personality, and the King's own doctor, Isaac Abrahams, brother of the Olympic sprinter Harold.

Wally got the impression that the doctors thought he was making up symptoms to get out of BD, but he was not. He had always felt that his life would be short, and the work gave him cause for anxiety: 'You was always wondering if you'd get back to the billet. But everyone was in a similar position, whether a civilian or in the

Forces. If you woke up the next day you considered yourself very lucky. I'm not what you call a religious fellow but I was brought up to go to church, and there was always a feeling that there was a supreme power protecting you.'

He was away for about three months, and decided that he did not want to go back to BD, where more and more time was being spent waiting at the billet for something to happen. He had attended a course on engineering and so transferred into the Royal Electrical and Mechanical Engineers. Yet he still had nightmares, even after leaving BD. Once he was woken up by the noise of a bomb, only to discover that it had been a goods train thundering through the station. The bad dreams continued into old age: the feeling of being down a hole with a UXB and wondering whether if it exploded it would miss him.

The scars of war have still to heal elsewhere, too. A century after the Great War, the Belgian armed forces turn up ten unexploded pieces of ordnance every day. Sixty years after the Second, Germany remains contaminated by tens of thousands of shells, mines and UXBs.

More than two thousand tons of wartime munitions are recovered each year in Germany, and it is a rare week that a street or a station is not closed to allow for disposal work. The worst-affected state is Brandenburg, around Berlin, which currently estimates that it will take another 150 years to clear up the vast amount of steel and explosive used against the city and later covered over by hasty post-war rebuilding.

At Christmas 1958 the reservoir of the Sorpe Dam was drained to begin repairing damage caused by the great raids on the Ruhr. At its base, sticking out of the mud, was the largest Second World War UXB ever found.

It was one of the twelve-thousand-pound Tallboys developed by Barnes Wallis for the RAF, used to sink the *Tirpitz* and dropped in the second attack on the dam, in October 1944. With a casing more than ten feet long and three feet wide, the weapon held 5200 pounds of Torpex high explosive and was capable of creating a crater eighty feet deep. As the RAF was able to tell the German bomb disposal team, it was fitted with three sensitive and potentially instable chemically operated pistols.

In early January 1959 the Tallboy became the focus of an Anglo-German BD effort. Led by Flight Lieutenant J. M. Waters and veteran *Feuerwerker* Walter Mitzke, the team was able to extract the first pistol remotely. The other two, however, needed to be retrieved by hand. This was done successfully, although it was subsequently found that in one of them the striker was touching the detonator but had not hit it hard enough to set off the warhead. Two more of the giant bombs that were subsequently found there were dealt with later in the month.

UXBs continue to be discovered wherever the conflict raged, notably in Japan and on islands in the Pacific. American BD squads on Guam have to deal with more than 250 reports every year. They are still found regularly in Britain as well. Following the broadcast of the television series *Danger UXB* in the late seventies, members of the public called in the Army to remove six butterfly bombs that had served as doorstops or household ornaments for some thirty years. In 1996 the government gave details of seventy-four known locations of UXBs in London alone – dozens more are thought to be buried deep in the Thames – while in 2008 a map to be distributed to building firms identified twenty-one thousand places in the UK where there might still be unexploded bombs.

During construction work on the site for the Olympic stadium in the summer of 2008, the largest UXB to be found in the capital for three decades was unearthed by a digger. It was a thousand-kilogram Hermann that had fallen on the banks of the River Lea near Three Mills Island. A two-hundred-yard exclusion zone was cordoned off and police considered evacuating all buildings within a mile; nearby London Underground and railway lines, which passed over a viaduct four hundred yards from the find, were closed.

Old bombs present their own problems. Those filled with powdered explosive containing aluminium powder and ammonium
nitrate can hold cavities that form when the charge deteriorates,
generating nitrogen oxide which can be set off by friction rasping
against iron picrates, the temperamental crystals bred by chemical
interaction with the metal surfaces of bombs.

The Bromley Hermann contained solid explosive, which
would need steaming out, but the chemicals in this could have
reacted to create a compound that detonates at well below the
temperature of steam. A section from 33 Engineer Regiment
(EOD) was called in. Working around the clock it erected a blast
wall around the bomb built from four hundred tonnes of sand.
Some things had not changed in seventy years, and a salt solution
was used to neutralise the fuze. Other things had: a laser-guided
water jet was used to trepan holes in the casing so that the explosive could be liquefied.

Although the equipment has improved vastly since the Second
World War, the spirit shown by British bomb disposal experts in
the conflicts since is recognisably the same. Be it in Korea,
Malaysia, Northern Ireland, the Falklands, the Balkans, or now in
Iraq and Afghanistan, the self-control, self-confidence and even
self-sacrifice demanded are no different from those shown when
UXBs were first grappled with during the Blitz. Many more
medals for gallantry have been won, including the George Crosses
awarded in 2006 to Major Peter Norton for his bomb disposal
work in Iraq, and in 2010 to Staff Sergeants Kim Hughes and Olaf
Schmid for their actions in Afghanistan. As shown by the posthumous nature of Schmid's GC, more casualties have also been taken
by those hailed by the Chief of the Defence Staff, Air Chief
Marshal Sir Jock Stirrup, as 'the bravest of the brave'. As long as
men feel the need to devise more ingenious and more deadly
methods of killing each other, there will be work for those willing to try to stop them.

THE WHOLE EARTH

On 14 May 2003 Queen Elizabeth II arrived at Westminster Abbey to keep faith with an act of national recognition begun by her father. Several years earlier Stuart Archer had become Chairman of the VC & GC Association, and had discovered that there was no public memorial to those who had won either medal.

He tried first to have a statue of Queen Victoria on her horse, pinning on the first VC, erected on the Fourth Plinth in Trafalgar Square but this was soon scotched. He was, however, able to arrange for a memorial stone to be placed in the floor of the west end of the Abbey. 'In my view,' he said proudly, 'it is greater than anything else I've ever done.'

The dedication of the stone was attended by eleven of the fifteen holders of the Victoria Cross still living, and by twenty-three of the twenty-nine surviving recipients of the George Cross, the decoration created at the height of the aerial bombardment of Britain. Before inviting the Queen to unveil the memorial, Archer read from Pericles's oration for the dead of another war, as recorded by Thucydides:

> And while committing to hope the uncertainty of final success, in the business before them they thought fit to act boldly and

trust in themselves. Thus choosing to die resisting, rather than to live submitting, they fled only from dishonour, but met danger face to face, and after one brief moment, while at the summit of their fortune, escaped, not from their fear, but from their glory . . .

For this offering of their lives made in common by them all they each of them individually received that renown which never grows old, and for a sepulchre, not so much that in which their bones have been deposited, but that noblest of shrines wherein their glory is laid up to be eternally remembered upon every occasion on which deed or story shall call for its commemoration. For heroes have the whole earth for their tomb.

DO's AND DONT's FOR BOMB DISPOSAL.

DO TAKE EVERY SAFETY PRECAUTION

DONT TAKE ANY UNNECESSARY RISKS.

DO REPORT ANYTHING STRANGE TO I.O.F. & D.B.D. (TEL. ABB. 2315) WHERE THE I.O. WILL SUPPLY ADVICE AND ASSISTANCE.

DONT TALK ABOUT YOUR EXPLOITS. IT MAY HELP THE ENEMY TO CHANGE HIS METHODS.

DO, WHEN WORKING ON A ⑰ FUZE, LISTEN AT FREQUENT INTERVALS (SAY 5 MINS) TO THE CLOCK.

DO READ AND DIGEST THE INTELLIGENCE SUMMARIES. THE INFORMATION MAY BE OF THE GREATEST USE TO YOU.

DONT LEAVE TOOLS AND PLANT LYING ABOUT OR THEY WILL BE MISSING NEXT TIME THEY ARE WANTED.

DO REMEMBER THAT IT IS YOUR DUTY TO INSPIRE CONFIDENCE, SO DO NOT TRY TO IMPRESS BY "TELLING THE TALE".

DONT TRY TO TAKE ANYTHING TO PIECES UNLESS YOU KNOW EXACTLY WHAT YOU ARE TRYING TO DO.

DONT ALLOW ANYONE BUT THE DRIVER ON A VEHICLE CARRYING A LIVE BOMB.

DONT FORGET - THAT THOUGH YOUR WORK IS INEVITABLY DIRTY - SMARTNESS COUNTS.

DO RESPECT PRIVATE PROPERTY - A BOMBED HOUSE IS STILL SOMEBODY'S HOME.

DONT COLLECT SOUVENIRS.

DO KEEP TOOLS & PLANT IN GOOD WORKING ORDER.

DONT LEAVE ESSENTIAL TOOLS AND STORES BEHIND.

DO KEEP ON CORDIAL TERMS WITH THE LOCAL CIVIL AUTHORITIES

AND REMEMBER ALWAYS THAT YOU BELONG TO THE CORPS OF

ROYAL ENGINEERS

BIBLIOGRAPHY

Books and unpublished memoirs

Adie, Kate, *Into Danger: Risking Your Life for Work* (Hodder, 2009)

Balchin, Nigel, *The Small Back Room* (Cassell & Co, 1943)

Bartlett, Vernon, *And Now, Tomorrow* (Chatto & Windus, 1960)

Beckingham, HW, *Living Dangerously with Bombs and Mines* (unpublished, 1989)

Bisset, Ian, *The George Cross* (MacGibbon & Kee, 1961)

Blank, Ralf et al (trans. Derry Cook-Radmore)), *Germany and the Second World War, Volume IX/I: German Wartime Society 1939–1945: Politicization, Disintegration and the Struggle for Survival* (Clarendon Press, 2008)

British Information Services, *Britain, Volume 1* (British Information Services, 1942)

Brown, Andrew, *J. D. Bernal: The Sage of Science* (Oxford University Press, 2007)

Calder, Angus, *The People's War: Britain 1935–45* (Granada, 1982)

———, *The Myth of the Blitz* (Jonathan Cape, 1991)

Churchill, Winston, *The Second World War, Volume II: Their Finest Hour* (Cassell & Co, 1949)

Clark, Ronald, *The Birth of the Bomb: The Untold Story of Britain's Part in the Weapon That Changed the World* (Phoenix House, 1961)

———, *The Greatest Power on Earth: The Story of Nuclear Fission* (Sidgwick & Jackson, 1980)

Colville, John, *The Fringes of Power: Downing Street Diaries 1939–1955* (Weidenfeld & Nicolson, 2004)

Crowther, J. G., and Whiddington, R., *Science at War* (HMSO, 1947)

De Courcy, Anne, *The Viceroy's Daughters: The Lives of the Curzon Sisters* (Weidenfeld & Nicolson, 2000)

Demarne, Cyril – *The London Blitz: A Fireman's Tale* (After the Battle, 1991)

Dilks, David (ed.), *The Diaries of Sir Alexander Cadogan, OM, 1938–1945* (Cassell & Co, 1971)

Edgerton, David, *Warfare State: Britain 1920–1970* (Cambridge University Press, 2006)

FitzGibbon, Constantine, *The Blitz* (Macdonald, 1970)

FitzGibbon, Theodora, *With Love* (Century, 1982)

Fleischer, Wolfgang, *German Air-Dropped Weapons to 1945* (Midland, 2004)

Frayn Turner, John, *The Blinding Flash: The Remarkable Story of Ken Revis and his Struggle to Overcome Blindness* (Harrap,1962)

Gardiner, Juliet, *Wartime: Britain 1939–1945* (Review, 2005)

Gaskin, M. J., *Blitz: The Story of 29th December 1940* (Faber and Faber, 2005)

Good, Irving John (ed.), *The Scientist Speculates: An Anthology of Partly Baked Ideas* (Heinemann, 1962)

Green, Henry, *Caught* (Hogarth Press, 1943)

Hebblethwaite, Marion, *One Step Further: Those Whose Gallantry was Rewarded with the George Cross, vols 1–9* (Chameleon HH Publishing, 2005–2010)

Hennessy, Peter, *Whitehall* (Fontana, 1990)

Hare-Scott, Kenneth, *For Gallantry: The George Cross* (Peter Garnett, 1951)

Hartley, A. B., *Unexploded Bomb: A History of Bomb Disposal* (Cassell & Co, 1958)

Hissey, Terry, *Come if Ye Dare: The Civil Defence George Crosses* (Civil Defence Association, 2008)

Hogben, Arthur, *Designed to Kill,* (Patrick Stephens, 1987)

Hunt, H. J., *Bombs and Booby Traps* (Romsey Medal Centre, 1986)

Jappy, M. J., *Danger UXB: The Remarkable Story of the Disposal of Unexploded Bombs during the Second World War* (Channel 4 Books, 2001)

Jeacocke, Leonard, *A Saturday Night Soldier* (unpublished, 1998)

Klemmer, Harvey, *They'll Never Quit* (Wilfrid Funk & Co, 1941)

Lacey, James, *103 Bomb Disposal Section: Diary of Events 27 June 1940 – 31 December 1940* (unpublished memoir, 1995)

Lofthouse, Alistair, *The Sheffield Blitz: Operation Crucible* (ALD Design & Print, 2001)

Longmate, Norman, *The Doodlebugs: The Story of the Flying-Bombs* (Hutchinson, 1981)

Lowe, Keith, *Inferno: The Devastation of Hamburg, 1943* (Penguin, 2007)

Macmillan, Harold, *The Blast of War, 1939–1945* (Macmillan, 1967)

Maier, Klaus et al (trans. Dean S. McMurray and Ewald Osers), *Germany and the Second World War, Volume II: Germany's Initial Conquests in Europe* (Clarendon Press, 1991)

Mason, Francis K., *Battle over Britain: A History of the German Air Assaults on Great Britain, 1917–18 and July–December 1940, and of the Development of Britain's Air Defences between the Wars* (McWhirter Twins Ltd, 1969)

Matthews, W. R., *Saint Paul's Cathedral in Wartime 1939–1945* (Hutchinson, 1946)

Minney, R. J. (ed.), *The War Weekly* (Newnes, 1939–1941)

Ministry of Home Security, *Front Line 1940–1941: The Official Story of the Civil Defence of Great Britain* (HMSO, 1942)

O'Brien, Terence H., *Civil Defence* (HMSO and Longmans, Green and Co, 1955)

Ovens, Rex, *Memoir* (unpublished, 1999)

Pakenham-Walsh, RP, *History of the Corps of Royal Engineers, vol. 8* (The Institution of Royal Engineers, 1958)

Panter-Downes, Mollie, *London War Notes, 1939–1945* (Longman, 1972)

Pile, Frederick, *Ack-Ack: Britain's Defence against Air Attack during the Second World War* (Harrap, 1949)

Printing & Stationery Services, *Summary of German Rheinmetall Fuzes* (Printing & Stationery Services, 1944)

Pyle, Ernie, *Ernie Pyle in England* (Robert M. McBride, New York, 1941)

Ramsey, Winston G. (ed.), *The Blitz Then and Now* (After the Battle, 1986)

Ransted, Chris, *Bomb Disposal and the British Casualties of WW2* (Chris Ransted, c. 2004)

Rose, Kenneth, *Elusive Rothschild: The Life of Victor, Third Baron* (Weidenfeld & Nicolson, 2003)

Rothschild, Victor, *Meditations of a Broomstick* (Collins, 1977)

Sansom, William, *The Blitz: Westminster at War* (Oxford University Press, 1990)

Scott, J. D., and Hughes, Richard, *The Administration of War Production* (HMSO, 1955)

Sheppard-Jones, Elisabeth, *I Walk on Wheels* (Geoffrey Bles, 1958)

Sinclair, Andrew, *The Red and the Blue: Intelligence, Treason and the Universities* (Weidenfeld & Nicolson, 1986)

Smyth, John, *The George Cross* (Arthur Barker Ltd, 1968)

Spick, Mike, *Luftwaffe Bomber Aces: Men, Machines, Methods* (Greenhill Books, 2001)

Strachey, John, *Post D: Some Experiences of an Air-Raid Warden* (Victor Gollancz, 1941)

Swann, Brenda, and Aprahamian, Francis (eds), *J. D. Bernal: A Life in Science and Politics* (Verso, 1999)

Wakeling, Eric, *The Danger of UXBs: True Stories of Bomb Disposal Heroism in World War II* (BD Publishing, 1996)

Wakeling, Eric, *The Lonely War: A Story of Bomb Disposal in World War II* (BD Publishing, 1998)

Wells, William, *Danger! Unexploded Bomb* (unpublished memoir, 1951, ed. 1998)

Ziegler, Philip, *London at War 1939-1945* (Sinclair-Stevenson, 1995)

Zuckerman, Solly, *From Apes to Warlords: The Autobiography* (Hamish Hamilton, 1978)

Articles

Christopherson, Derman, and Baughan, E.C, 'Reminiscences of Operational Research in World War II by some of its practitioners: II', *Journal of the Operational Research Society*, vol. 43, no. 6, June 1992

Cook, A. H., 'Ian Morris Heilbron', *Biographical Memoirs of Fellows of the Royal Society*, vol. 6, November 1960

Cottrell, Alan, 'Edward Neville da Costa Andrade', *Biographical Memoirs of Fellows of the Royal Society* vol. 18, November 1972

Dorey, S. F., 'Herbert Gough', *Biographical Memoirs of Fellows of the Royal Society* vol. 12, November 1966

Gough, H. J., 'Research and Development Applied to Bomb Disposal:

the Thirty-Third Thomas Hawksley Lecture', *Proceedings of the Institution of Mechanical Engineers*, 156

Ransted, Chris, 'The Death of an Earl', *After the Battle*, no. 146

Sargant, William, and Craske, Nellie, 'Modified Insulin Therapy in War Neuroses', *The Lancet*, vol. 238, issue 6156, 23 August 1941

GERMAN EXPLOSIVE ORDNANCE

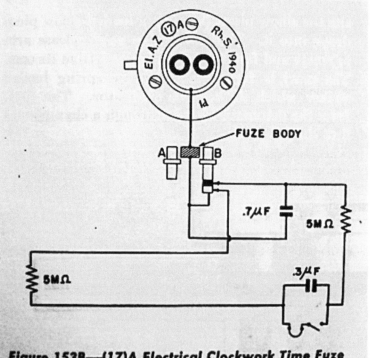

Figure 153B—(17)A Electrical Clockwork Time Fuze Wiring Diagram

GLOSSARY

Fuze and bomb parts

main charge or filling: the explosive in the body of the bomb
condenser: a type of battery that holds electric charge
fuze pocket: cylindrical container that holds the fuze and picrics
fuze boss: the top face of the fuze, stamped with its fuze number
gaine: initiating charge set off by the fuze and screwed into its base
picric pellets: secondary booster charge under the gaine

Common types of German fuze

15 and 25: explodes on or very soon after impact
17: long delay, explodes up to four days after impact
28 and 38: very short delay, used mainly against naval targets
50: motion-sensitive anti-handling fuze
Y: a fuze intended to kill bomb disposers, triggered by movement in any
     direction
Zus 40: booby trap designed to prevent removal of (17) fuze

Categories of UXB

A1: immediate disposal essential, detonation *in situ* not acceptable
A2: immediate disposal essential, detonation *in situ* permissible
B: urgent disposal required
C: not necessarily calling for immediate disposal
D: To be disposed of as convenient

NOTES

Abbreviations

NA   National Archives, Kew

IWM   Imperial War Museum (SA: Sound Archive; DD: Department of Documents)

EODTIC   Explosive Ordnance Disposal Technical Information Centre

REMLA   Royal Engineers Museum, Library and Archive

AIME   Archive of the Institution of Mechanical Engineers

JHA   Papers of Professor John Hudson

Foreword: A Nation on Guard

xi   *broadcast by their king*: The text of the King's speech can be found in *Flight* magazine, 3 October 1940, p. 265.

xiv   *'A bomb is a pretty lethal thing'*: John Hannaford cited in Trevor Royle, 'A most disarming game of cat and mouse', *Sunday Herald*, 11 February 2001.

xv   *'Somehow or other'*: Churchill, *Their Finest Hour*, p. 362.

1: Danger UXB

2   *'a sticky one'*: Jeacocke, *A Saturday Night Soldier*, p. 17.

7   *In the evidence*: I have reconstructed the Wanstead episode from several sources: (i) Herbert Hunt's bomb diary [REMLA 9511.9.13]; (ii) the Company war diary [NA WO 166/3997]; (iii) Bill Hartley's account in *Unexploded Bomb*, p. 53, which seems to draw on the

testimony of a survivor, probably Fox; and (iv) Len Jeacocke's reminiscences in *A Saturday Night Soldier*, p. 8. The first two state that
the UXB had only a (17) time fuze – Hunt, who was CO of
another section in the area, claims it had two – and neither mention a (50). Since they are the more contemporaneous sources, I
would normally give them most weight, but Hartley's version has
so much specific detail that it is difficult to discount it. Accordingly,
I have preferred that, supported as it is by Jeacocke's recollection
that there was a (50) present. The recommendation and citation for
Blaney's GC at NA T 351/6 and WO 373/66 appear to be, at the
least, ambiguous regarding some details of the incident. This has
led to the inaccurate report that Blaney was dealing with the bomb
by himself. For biographical detail on Blaney see Hebblethwaite,
One Step Further, vol. 2, pp. 84–90.

8   *more than two hundred*: NA PREM 3/75/1, 3.1.41. This figure
does not include those BD casualties sustained to date by the other
Armed Services.

2: The Shape of Things to Come

10   *One warm afternoon*: Jonathan Keates, *The Siege of Venice* (Chatto &
Windus, 2005), p. 385.
11   *'a wildly hazardous procedure'*: Hartley, *Unexploded Bomb*, p. 2.
15   *'none of the major four components'*: Rühlemann, *Father Tells Daughter*,
p. 116.
16   *'our design evaluation'*: Ibid., p. 136.
18   *'Bombs dropped from aeroplanes'*: Ibid., p. 158.
19   *'Secrecy was the key word'*: Ibid., p. 153.
20   *'We had not one failure'*: Ibid., p. 165.
20   *'It could ignite'*: Ibid., p. 225.
21   *'I was at no time'*: Ibid., p. vi.
21   *now £225,000*: I have used the Economic History Association's
website www.measuringworth.com to calculate the comparative
figures.
22   *'The starting point'*: Christopherson and Baughan, 'Reminiscences
of Operational Research in World War II by some of its
Practitioners'.
22   *sixty-six thousand people*: O'Brien, *Civil Defence*, pp. 96 and 144.
23   *'The Ministry of Some Obscurity'*: Ibid., p. 302n.

24   *So it was*: Pakenham-Walsh, *History of the Corps of Royal Engingeers*, vol. 8, p. 123, and Hunt, *Bombs and Booby Traps*, p. 121.

24   *'at the end of which'*: Beckingham, *Living Dangerously*, p. 8, in Private Papers of H. W. Beckingham, IWM DD 90/29/1.

25   *'we languished'*: Ibid.

25   *'The scheme is ingenious'*: NA AVIA 13/507.

26   *'I couldn't argue with this'*: IWM SA 6828

26   *71,000 tons*: O'Brien, op. cit., p. 680.

27   *Arthur Merriman*: The biographical details are taken chiefly from Hissey, *Come if Ye Dare*, pp. 20–6.

27   *What the pair found*: NA AVIA 22/2263. Merriman's report is at tab 11A.

29   *'A condition may arise'*: Ibid.

30   *First tool specifically invented*: NA WO 195/104.

30   *Len Harrison*: For Harrison and the Immingham and Grimsby UXBs, see IWM SA 6828.

32   *'As soon as you introduce'*: Rühlemann, op. cit., p. 385.

33   *'several hundred feet wide'*: Ibid., p. 303.

3: Wars are Won in Whitehall

34   *On a warm evening*: For the Tots and their dinner, see: Zuckerman, *From Apes to Warlords*, p. 109; Brown, *J. D. Bernal – The Sage of Science*, p. 176; Swann and Aprahamian, *J. D. Bernal*, p. 168.

36   *Herbert Gough*: For Gough's life hitherto, see in particular *Dictionary of National Biography* (101033488) and his proposal for membership of the IME, AIME, 1933. He did move to the Ministry of Supply as in 1939, not 1942 as implied by *Who Was Who* and stated in the DNB: viz. *British Imperial Calendar and Service List 1940*, p. 378.

37   *'a conscientious worker … keen and critical mind'*: 'Herbert Gough', *Biographical Memoirs of Fellows of the Royal Society*, vol. 12, p. 190.

38   *'First, service in such a post'*: NA SUPP 20/5, 1A.

39   *'his work was his greatest joy'*: AIME COP 18/6.

39   *'at the best gifted amateurs'*: Scott and Hughes, *The Administration of War Production*, p. 270.

40   *'This, at any rate'*: Ibid., p. 272. See also Edgerton, *Warfare State*, p. 160 et seq. for an outline of the relations between the military and scientific personnel at the Ministry.

41   *'a clear statement'*: NA AVIA 22/2278, 14.5.40.

41 *'most strenuous and complicated'*: Hansard, House of Commons Debate, 27 June 1940, vol. 362, cc. 623–4. At 627, Morrison explains some of the reasons for the tank debacle.

41 *'My outstanding impression'*: 'Herbert Gough', op. cit., p. 187.

42 *'necessary further research'*: NA AVIA 22/2263, 5.4.40.

42 *'No details'*: Ibid., 4C Appendix A.

42 *'the handling of an unexploded bomb'*: NA WO 195/104.

42 *'in a bath of liquid air'*: Ibid.

43 *'not encouraging'*: Ibid.

43 *'I can still remember'*: J. D. Bernal in Good (ed.), *The Scientist Speculates*, p. 11.

43 *'most promising method'*: NA WO 195/104.

43 *'mechanical device'*: Ibid.

44 *'We were a group'*: Rühlemann, *Father Tells Daughter*, p. 351.

45 *Ian Heilbron*: For Heilbron's biography, see especially *Dictionary of National Biography* (101033799) and *Biographical Memoirs of Fellows of the Royal Society*, vol. 6, pp. 65–85.

46 *Edward Andrade*: For Andrade, see *Biographical Memoirs of Fellows of the Royal Society*, vol. 18, pp. 1–20.

47 *'to co-ordinate experimentation'*: NA AVIA 22/2263, 14.2.40.

47 *'to consider the general problem'*: EODTIC UK/ H8-1 E3, for the original minutes of the 1 May meeting. See NA WO 195/170 for a summary of them. The National Archives holds almost six hundred reports on the work of the Unexploded Bomb Committee and its sub-committees. See also, Gough to Lord Cadman, NA AVIA 22/2263, asking for authority to set up such a sub-committee of the Scientific Advisory Council.

48 *Two and a half thousand*: NA SUPP 4/251.

49 *250 fuze extractors*: Ibid.

4: I May Not Last Very Long at This

51 *'A convenient title'*: NA AVIA 22/2263, 7B.

51 *'i.e. a digging party'*: Ibid.

52 *'personnel for these sections'*: EODTIC UK / H 8-25, 26.5.40.

52 *'They were given'*: IWM SA 6828.

53 *'We found the aerodrome'*: Wells, *Danger! Unexploded Bomb*, p. 7, in Private Papers of W. H. Wells, IWM DD 12479.

53 *'Nobody seemed to worry'*: Ibid.

54 *Stuart Archer*: For Stuart Archer's early life the main sources are: Hebblethwaite, *One Step Further*, vol. 1, p. 45; IWM DD A 13, GC Box 1; IWM SA 25426; and interview with Stuart Archer by the author, 1 July 2008.

55 'in licensed premises': IWM SA 25426.

55 'One could appreciate': Interview, 1 July 2008.

55 'the right and decent thing': IWM SA 25426.

55 'They spoke very loudly': Ibid.

56 'It seemed to me': Ibid. Also, interview 1 July 2008.

56 'rather surprised': Adie, *Into Danger*, p. 191.

56 'They were trying to wave me': IWM SA 25426.

57 'we had, we all thought': Ibid.

57 'They didn't last': Ibid.

57 'They already had': Ibid.

57 'They were looking round': Ibid.

57 'This ... is a German bomb': Ibid.

58 'At that time': Ibid.

58 'I think really people': Interview, 1 July 2008.

58 'I got on extraordinarily well with him': IWM SA 25426.

58 'You're Bomb Disposal': Ibid.

59 'It was about fifteen inches': Ibid.

5: Wild Jack

60 *smart flannel trousers*: For the scene aboard *Broompark*, see: von Halban's account in Clark, *The Birth of the Bomb*, p. 99; Clark, *The Greatest Power on Earth*, p. 98; and Macmillan, *The Blast of War*, p. 100.

61 *Wars create legends*: The main sources for Suffolk's early life are my interviews with his sons, the Earl of Suffolk, 28–29 June 2008, and the Hon Maurice Howard, 28–29 June 2008 and 17 September 2009. See also: Hare-Scott, *For Gallantry*, pp. 104–8; Smyth, *The George Cross*, pp. 28–30; Bissett, *The George Cross*, p. 56; Hissey, *Come if Ye Dare*, pp. 58–60; Hebblethwaite, *One Step Further*, vol. 8, pp. 71–2; and IWM DD GC Box 14.

63 'entered freely into the life': *Courier Mail*, 23 February 1934, p. 13.

64 'They were a couple': Interview, 16 July 2008

64 'anyone on the staff': Ibid.

64 'charm of manner': Hare-Scott, op. cit., p. 107.

65   *'I was frightened'*: Ibid., p. 108.

65   *A stream of letters*: NA AVIA 22/2288A.

66   *guide Desmond Bernal around Paris*: Zuckerman, *From Apes to Warlords*, p. 116.

66   *worrying unnecessarily*: NA AVIA 22/2288A.

67   *'the most utter chaos'*: NA AVIA22/3201.

67   *'upon our own initiative'*: Ibid.

67   *'with nothing but'*: Ibid.

67   *'in the most uncompromising'*: Ibid.

68   *'I put my wife'*: Clark, *Birth of the Bomb*, p. 95.

69   *'By Tuesday evening'*: NA AVIA 22/3201.

69   *'faites en mer'*: NA AVIA 22/2288A.

69   *'a young man'*: Macmillan, op. cit., p. 100. Much of the remainder of Macmillan's account of the exploit, and the article of May 1943 in *Reader's Digest* by Suffolk's brother, Greville Howard, on which it is based, has no basis in fact. Golding wrote to the *Sunday Times* on 27 March 1960 to correct some of the exaggerations, such as the story that *Broompark* had repeatedly been dive-bombed.

70   *'I have never known'*: Macmillan, op. cit., p. 103.

70   *'explosive in the ordinary sense'*: NA AVIA 22/2288A.

70   *'my scientific friends'*: Ibid.

71   *'I have a very unusual request'*: Ibid.

71   *'The King knows it is here'*: Ibid.

71   *'some of them'*: NA AVIA 22/3201.

72   *'brilliantly handled'*: *The Times*, 21 May 1941.

72   *'expounded with boyish delight'*: Ibid.

6: Something Wicked ...

74   *Report from A.E.*: Dilks (ed.), *The Diaries of Sir Alexander Cadogan*, p. 308.

74   *'He was a signwriter'*: IWM SA 25426.

74   *'We were very much'*: Ibid.

75   *'The demand for clothes'*: Lacey, *103 Bomb Disposal Section*, REMLA, 9503.6.

75   *'With the thought'*: Ibid.

75   *'We were just'*: Ibid.

76   *'bed was three blankets'*: Ibid.

76   *'The wardens didn't know'*: IWM SA 25426.

76   *'We knew later'*: Ibid.

77   *'We didn't attempt'*: Ibid.

77   *'I became aware'*: Ibid.

77   *'I think I accepted'*: Ibid.

77   *'that these young men'*: Ibid.

77   *'so presumably'*: Ibid.

79   *'We were rather dubious'*: Wells, *Danger! Unexploded Bomb*, p. 9.

80   *In daylight*: Spick, *Luftwaffe Bomber Aces*, p. 113.

81   *'1. The bombs are dropped too low'*: Fleischer, *German Air-Dropped Weapons to 1945*, p. 72.

81   *A report would show*: This was the product of David Benusson-Butt's forensic examination of the reconnaissance photographs of the bombing raids.

81   *On the morning*: *London Gazette*, 17 January 1940 and *The Times*, 18 September 1940. See also, Hebblethwaite, *One Step Further*, vol. 3, pp. 148–9.

83   *'would cause disruption'*: Jappy, *Danger UXB*, p. 43.

84   *'For the moment'*: NA WO 195/83.

84   *'as usual'*: Letter in Hebblethwaite, op. cit., vol. 6, p. 154.

84   *In Swansea*: Lacey, op. cit.

85   *spindles and cogwheels*: IWM SA 23212.

85   *One such occurred*: *London Gazette*, 17 December 1940 and 28 October 1941.

86   *On 24 August*: For the Loughor UXBs, see *London Gazette*, 17 September 1940, and Lacey, op. cit.

87   *'piggy in the middle'*: Eric Wakeling in Royle, 'A most disarming game of cat and mouse'. Also IWM SA 26831.

7: ... This Way Comes

89   *'a lot of donkey work'*: IWM SA 25426.

89   *' safe as houses'*: Ibid.

89   *St Athan aerodrome*: For this exploit, that at Moulton and the others detailed in the recommendation and citation for Archer's George Cross, see NA T 351/9 and WO 373/66. Archer says in IWM SA 25426 that the fuze used at St Athan was a (17), but this seems improbable given that it was a month before Mitchell made his discovery of what is accepted to be the first time fuze.

90   *' hell of a chance'*: IWM SA 25426.

90 'It sounds a bit': Ibid.
90 'If you imagine': Ibid.
90 It was similar in principle: EODTIC H/GE/24, enclosure 31.
91 'It should have gone off': IWM SA 25426.
92 'delay action bombs': NA CAB 120/266, 25.8.40.
93 The very next day: EODTIC UK/H8-26.
94 'matters of equipment and research': NA AVIA 22/2263, 6.9.40.
96 'with undaunted and unfailing courage': London Gazette, 21 January 1941.
96 Difficulties with getting it into service: NA WO 195/296.
96 steam apparatus: NA WO 195/301.
97 meeting of the Board: NA AVIA 22/2263, 26.8.40.
98 He managed to remove: NA WO 373/66 Pt 1.
98 In Moxey's Wolseley: Hebblethwaite, One Step Further, vol. 6, p. 151.
99 The dual attack: NA CAB 68/7/18.
99 When they arrived: For the outline of Archer's exploit at Llandarcy, see NA WO 373/66 and IWM SA 25426. The extent of the damage to the refinery is detailed in NA CAB 66/11/41 and NA CAB 68/7/18.
100 slid down head first: Interview with Deirdre Strowger, 25 November 2009.
100 'So then': IWM SA 25426.
101 'So I grabbed': Ibid.
101 'I looked inside': Ibid.
101 very simple principles: See the technical examination in NA WO 195/1286.
102 Meanwhile, at Great Livermere: NA T351/1 and interview with John Emlyn Jones, 26 January 2010. The citation for Emlyn Jones's MBE gives the date of the extraction of this fuze as 31 August. If correct, it may mean that he rather than Archer was the first to recover a Zus, despite the wording of Archer's citation. Emlyn Jones remembers taking it to Captain Kennedy at the War Office, although on which day is unclear.
102 'When you are sitting': Sunday Telegraph, 4 November 2001.
103 'I take my hat off': Ibid.
103 'We had a young officer': Ibid.
103 'It was extremely difficult for her': Ibid. Also in IWM SA 25426.

8: Red Sky at Night

104 *The fires at the docks*: See MOHS, *Front Line 1940–1941*, especially pp. 11–12 and 25.

105 *'Most of us'*: Private Papers of W. C. Bowman, IWM DD 3769.

106 *'Great canopies of billowing smoke'*: Private Papers of W. B. Regan, IWM DD 781.

106 *'fierce crackle'*: Ibid.

106 *'they were screaming'*: Quoted in Ziegler, *London at War 1939–1945*, p. 121.

106 *'because once you've gone'*: Ibid. Celia Fremlin later became a writer of mystery stories.

106 *'no talk of horrors'*: Private Papers of W. C. Bowman, IWM DD 3769.

107 *'It was not till later'*: FitzGibbon, *The Blitz*, p. 118.

107 *The Luftwaffe defined*: O'Brien, *Civil Defence*, p. 681n.

107 *the Nazis' planners*: Maier et al., *Germany and the Second World War*, vol. 2, p. 390.

107 *The Port of London*: Gardiner, *Wartime*, p. 331.

107 *More than fifty thousand*: O'Brien, op. cit., p. 680.

108 *In the middle of the street*: See Calder, *The People's War*, p. 198.

108 *The bodies identified*: Ibid., p. 233.

108 *'to be sure to have one's handbag'*: Private Papers of M. E. Allan, IWM DD 3100.

108 *The first (17)*: Sansom, *The Blitz*, p. 28.

109 *gold coins*: Ibid., p. 35.

109 *'She showed me'*: Private Papers of W. B. Regan, IWM DD 781.

109 *Just twelve sections*: NA WO 166/3997.

109 *'most of the trouble'*: NA PREM 3/75/1, 11.9.40.

109 *'new and damaging'*: Churchill, *Their Finest Hour*, p. 360.

110 *'expand as rapidly'*: NA PREM 3/75/1, 13.9.40.

110 *'a task of the utmost peril'*: Churchill, op. cit., p. 360.

110 *'It wasn't a very popular piece'*: Wells, *Danger! Unexploded Bomb*, p. 9.

111 *'I shall not easily forget'*: Hare-Scott, *For Gallantry*, p. 115.

113 *Outside the cordon*: Sansom, op. cit., p. 33.

9: Saving St Paul's

114 *'A story that must win'*: *Daily Express*, 16 September 1940.

114 *Bob Davies was almost forty*: For Davies's outline biographical details, see: IWM DA GC Box 4; Hebblethwaite, *One Step Further*, vol. 3,

pp. 89–92; Army Service Record (No. 122933), Army Personnel Centre; Canadian service file (No. 2768756), Library and Archives Canada.

115 *Although he lied*: Attestation papers, service file, op. cit., Library and Archives Canada.

115 *a construction job*: *Daily Express*, 18 September 1940.

115 *When it was decided*: See 5 BD Company's War Diary, September 1940 – September 1941, at NA WO 166/3997.

116 *Then, as Davies hurried*: Ibid.

116 *When Davies arrived*: The fullest account of the hunt for the UXB is in Matthews, *Saint Paul's Cathedral in Wartime 1939–1945*, pp. 36–7.

116 *'For some reason'*: Ibid., p. 36.

117 *Nearby was a party*: *The Times*, 18 September 1940, and NA WO 166/3997.

118 *'These most gallant … men'*: *Daily Mail*, 13 September 1940.

118 *metal on metal*: *Daily Mail*, 16 September 1940.

118 *'a vast hog'*: *Daily Express*, 17 September 1940.

119 *'London is wild with praise'*: www.bbc.co.uk/ww2peopleswar/stories/88/a3295488.shtml, 17.9.40.

119 *'You couldn't have a better lot'*: *Daily Mail*, 17 September 1940.

119 *'Scared?'*: Ibid.

119 *'suicide squad'*: *Daily Mirror*, 17 September 1940.

119 *'saved the cathedral'*: *Daily Mail*, 16 September 1940.

119 *'I don't mind being gassed'*: *Daily Mail*, 17 September 1940.

119 *Thirty-one years of age*: For Wyllie's biography, see: Hebblethwaite, op. cit., vol. 9, pp. 68–70; his Army Service Record (No. 1945231), Army Personnel Centre; and *Glasgow Herald*, 4 October 1984.

120 *'When I found out'*: *Glasgow Herald*, 4 October 1984.

120 *'only the courage'*: Cited by *The Times*, 16 September 1940.

120 *'devotion to duty'*: Ibid.

120 *'the outstanding deed'*: *The Times*, 17 September 1940.

120 *'It is just like Bob'*: *Daily Express*, 18 September 1940.

121 *'It was just in his line'*: *Daily Express*, 17 September 1940.

121 *'cold-blooded disregard'*: *The Times*, 17 September 1940.

10: Run, Rabbit, Run

123 *By the end of November*: MOHS, *Front Line 1940–1941*, p. 61.

123 *fifty thousand houses*: Ibid., p. 71.

123 *'Ralph's beautiful antique desk'*: Private Papers of Glwadys Cox, IWM DD 2769.

124 *5500 in the space of three weeks*: NA CAB 68/7/16. This is the Civil Defence report for September 1940.

124 *A shortage of cement*: Calder, *The People's War*, p. 208.

124 *much resentment*: *Daily Express*, 17 September 1940.

125 *Plans were made*: *Daily Express*, 20 September 1940.

125 *Official sanction*: 'Reminiscences of Operational Research in World War II by some of its Practitioners'.

125 *'Many of these people'*: Pyle, *Ernie Pyle in England*, p. 114.

126 *as many as two-thirds*: MOHS, op. cit., p. 61.

126 *'It is true and comforting'*: Strachey, *Post D*, p. 67.

126 *'does not like his neighbours'*: C. FitzGibbon, *The Blitz*, p. 114.

126 *'People were concerned'*: T. FitzGibbon, *With Love*, p. 63.

127 *'George called me'*: Private Papers of W. B. Regan, IWM DD 781.

127 *'I looked around'*: Ibid.

128 *'I was unconscious'*: IWM SA 9878.

128 *'But my feeling'*: Ibid.

129 *'She was crying'*: IWM SA 18258.

129 *'I went straight'*: Ibid.

129 *'I admit I found'*: British Information Services, *Britain, Volume 1*, p. 82.

130 *From 10 September*: Pile, *Ack-Ack*, p. 152.

131 *Pieces of the bomb*: NA WO 166/3997.

132 *'We got there on Sunday'*: IWM SA 14840.

132 *'The best thing'*: Ovens, *Memoir*, p. 47.

133 *'magic words'*: Wells, *Danger! Unexploded Bomb*, p. 34.

134 *'as the happiest'*: IWM SA 14840.

134 *The first problem*: Frayn Turner, *The Blinding Flash*, p. 36.

135 *'There was very little'*: IWM SA 26831.

135 *It was not a perception*: Fourteen officers were killed in the five months of BD work up to 1 January 1941. By then, there were about 190 RE officers engaged on BD work sections, and 5900 men (plus 140 officers and 3900 troops from quarrying and construction companies). In that time, 110 sappers of other ranks died while on BD duty. The ratio of those killed to wounded by UXB

explosions was usually about two to one.

135 *'I think you found out'*: Ibid.
136 *'he had had'*: IWM SA 30828.
137 *'This lady had left'*: Jappy, *Danger UXB*, p. 81.
137 *'It came as no surprise'*: IWM SA 14158.
138 *'It blew the corporal'*: Ibid.
138 *'I had no idea'*: Jappy, op. cit., p. 82.
138 *'We started one day'*: IWM SA 14158.
139 *Len Jeacocke*: Interview with Barbara Jeacocke, 29 November 2009, and Jeacocke, *A Saturday Night Soldier.*
139 *'Officially, only officers'*: IWM SA 5339.
139 *'Two Welsh ex-miners'*: *TV Times*, January 1979.
140 *'Don't worry, boss'*: Ovens, op. cit., p. 54.
140 *'I used to say'*: IWM SA 14158.
141 *'Some blokes'*: *TV Times*, op. cit.
142 *'horrified to see'*: Wells, op. cit., p. 28.
143 *'The rapid disposal'*: NA CAB 120/266, 20.9.40.
144 *'The essence of this business'*: NA CAB 120/266, 14.9.40.
144 *'DSR. Very urgent'*: JHA. NA AVIA 22/2263, 23.9.40.
144 *'everyone is fully alive'*: NA CAB 120/266, 27.9.40.
144 *'We are alive'*: NA CAB 120/266, 7.10.40.
145 *'definitely safe to handle'*: NA WO 195/338. This is the survey by Merriman in late September 1940 of the fuzes encountered and the methods used against them.
145 *Trials were also ongoing*: Ibid.
145 *Major Ken Merrylees*: Obituary, *Daily Telegraph*, 20 July 1994.
146 *A series of six tests*: NA WO 195/357.
146 *'The only proper scientific deduction'*: Ibid.
146 *Frank Martin*: NA WO 166/4004 and WO 373/66.
147 *Herbert Hunt*: See Hunt, *Bombs and Booby Traps.*
150 *'I understood how my wife'*: Ibid., p. 110.
150 *'to contact a friend'*: Ibid.
151 *'I felt like exploding'*: Wells, op. cit., p. 32.

11: For Gallantry

152 *a 'premature bomb explosion'*: NA WO 166/3997.
152 *Somebody turned on the radio*: IWM SA 2806.
152 *The original impulse*: NA PREM 2/104.

153　*'Home Defence Medal'*: NA HO 186/2908, 14.9.40.

153　*'the civil population'*: NA HO 186/2908, 17.9.40.

153　*'At the moment'*: NA PREM 2/104, 18.9.40.

153　*'I thought we might'*: NA PREM 2/104., 17.9.40.

153　*agreed by the Cabinet*: NA HO 186/2908, 17.9.40.

153　*'A man the world'*: Daily Express, 21 September 1940.

154　*the news broke*: London Gazette (Issue 34956, supplement of 30 September 1940), 27 September 1940. The gasmen who had helped the squad later received £2 10s each from the Dean and Chapter of St Paul's as a token of their thanks: The Times, 27 December 1940.

154　*'Well, I reckon'*: Daily Mirror, 2 October 1940.

154　*'It seems as if'*: Ibid.

155　*'Thank God for that!'*: Interview with Stuart Archer, 1 July 2008.

155　*could form a crater*: Fleischer, German Air-Dropped Weapons to 1945, p. 10.

156　*The best test*: Notes and pamphlets in JHA.

156　*'We have not heard'*: NA CAB 120/266, 9.10.40.

157　*In the company*: NA WO 166/4000.

157　*Another measure*: NA CAB 120/266, 11.10.40.

157　*8.5 per cent*: Gough, 'Research and Development Applied to Bomb Disposal', p. 8. For more statistics on numbers of UXBs, see also EODTIC UK /H8/26 Part 2, E 23.

157　*Divisional Officers*: Hunt, Bombs and Booby Traps, p. 47 summarises their work.

158　*'right and proper'*: Wells, Danger! Unexploded Bomb, p. 26.

159　*Two sappers*: NA 166/4000 and Lacey, 103 Bomb Disposal Section.

159　*'Being no different'*: IWM DA 9084 948/1.

159　*'I think'*: Wells, op. cit., p. 25.

160　*As befitted a young gentleman*: The Times, 9 November 1940. See also Klemmer, They'll Never Quit, p. 103.

160　*'a groggy heart'*: Time, 18 November 1940.

161　*accused of stealing*: NA WO 166/3997.

161　*a pair of special constables*: East Anglian Daily Times, 1 October 1940.

161　*to 'listen in to delayed-action bombs'*: The Times, 5 November 1940.

161　*Victor Langston*: The Times, 12 November 1940.

162　*'After the fuze was removed'*: The War, p. 1268.

162　*'dearth of Section records'*: NA WO 166/3997.

162　*Lionel Carter*: Hunt, op. cit., p. 15. For other details of this episode, see Jappy, *Danger UXB*, pp. 29 and 79, for Brinton's account, and NA WO 166/3997.

163 *The clocks of the mine's fuzes*: For a history of the mines, see in particular Hogben, *Designed to Kill*, p. 31 et seq.

164 *the watching bomber crews*: Fleischer, op. cit., p. 70.

164 *'just as though'*: Sansom, *The Blitz*, p. 78.

166 *Jack Easton*: See Hebblethwaite, *One Step Further*, vol. 3, pp. 132–4.

166 *To their right*: Quoted in Ibid., p. 134. See also NA T 351/1.

167 *'Unless I got clear'*: Hebblethwaite, op. cit., p. 134.

167 *'I heard no explosion'*: Ibid.

167 *An hour later*: NA WO 166/3997.

168 *five hundred pounds*: The Times, 19 May 1942.

168 *at 1800 kilograms*: NA WO 166/3997.

169 *a third of a million tons*: Sansom, op. cit., p. 148.

169 *bulldozers*: The War, p. 1623.

169 *Bert Woolhouse*: IWM SA 20693.

170 *'He was only a short fellow'*: Ibid.

170 *'he was fearless'*: Ibid.

170 *'When you're on'*: Ibid.

171 *'And five pints'*: Ibid.

172 *'Titchie and Jackie'*: Ibid.

172 *five men had been killed*: Barking & Dagenham Post, 8 December 2008.

173 *The cause of the blast*: Hunt's notes and Wild's statement are at REMLA 9511.9.13-16. See also Hunt, op. cit., p. 100.

173 *Ash had dealt*: REMLA 9511.9.13.

173 *a recommendation*: EODTIC UK/H8-25.

173 *When Jane Woolhouse*: IWM SA 20694.

174 *'He'd kept from me'*: Ibid.

174 *'I kept shaking'*: IWM SA 20693.

174 *'a bloody madhouse'*: Ibid.

174 *'anxious, depressive or hysterical symptoms'*: See Sargant and Craske, 'Modified Insulin Therapy in War Neuroses'.

175 *'I can't help it'*: IWM SA 20693.

175 *'it was very difficult'*: IWM SA 20694.

12: Still Falls the Rain

176 *'All of a sudden'*: IWM SA 20693.

176 *an Anthony Eden*: A type of felt hat associated with (though not

often worn by) the statesman when a young and dandyish Foreign Secretary in the mid-thirties.

176 *'extremely violently Bolshevistic'*: NA AVIA 22/2288B.

176 *In late summer*: NA AVIA 22/2263, 11.9.40.

177 *The first was the formation*: NA AVIA 22/2263, 4.10.40.

177 *'a very serious gap'*: NA AVIA 22/2263, 22.10.40.

178 *'At one moment'*: Zuckerman, *From Apes to Warlords*, p. 128.

178 *Suffolk had first developed*: Hare-Scott, *For Gallantry*, p. 109.

178 *By 7 October*: NA AVIA 22/2263, 24.10.40.

178 *More remote areas*: Mentioned by Hudson, IWM SA 23212.

179 *'perceptible interval'*: Brown, *J. D Bernal*, p. 180 (from *New Scientist*, vol. 97, 1983, p. 904.)

179 *A photograph taken at the time*: Part of it appeared in the *Daily Sketch*, 19 July 1941.

179 *'I asked him'*: Parliamentary Debates, House of Lords, Official Report vol. 119, no. 43, Tuesday 20 May 1941, p. 197.

180 *'quite fearless'*: NA AVIA 22/1221.

180 *'Never mind, Fredders'*: Interview with June Daughtrey, 23 March 2010.

180 *'within approximately'*: NA AVIA 22/2254.

181 *'as a reliable means'*: NA AVIA 22/2263.

181 *'no other practicable course'*: NA AVIA 22/1221, 31.5.41.

181 *But fuzes had to be*: Ibid.

181 *Other exploits*: NA AVIA 22/2254.

182 *'only about 80 per cent safe'*: NA AVIA 22/2457.

182 *A carbine*: EODTIC UK/H/GE/29/1 (Basset and Hubbard report, p. 20).

182 *In its first report*: NA WO 195/441.

183 *Attempts to drill*: NA WO 195/414.

183 *a weak spot*: Gough, 'Research and Development Applied to Bomb Disposal', and NA WO 195/386.

183 *blow from a sledgehammer*: Hartley, *Unexploded Bomb*, p. 257.

184 *a change in strategy*: Maier et al, *Germany and the Second World War*, vol. 2, p. 399.

185 *In the first few weeks*: MOHS, *Front Line 1940-1941*, p. 19.

185 *Its sheer size*: Calder, *The People's War*, p. 237.

186 *'There were more open signs'*: Ibid., p. 204.

186 *The performance of the provincial fire brigades*: Ibid., p. 241.

187 *On 28 October*: London *Gazette*, 22 January 1941.

187 *The task fell*: The recommendations for Campbell and Gibson's GCs are at NA T 351/1 and WO 373/66 respectively. The biographical details come from Hebblethwaite, *One Step Further*, vol. 3, p. 13, for Campbell and supplement 2, p. 35, for Gibson, as well as the Newcastle *Evening Chronicle*, 13 September 2008.

188 *It was thought afterwards*: JHA, casualty analysis.

189 *On the night of 12 December*: The details of John Hudson's time in Sheffield are taken from IWM SA 23212 and the material in his archive.

189 *'My transport was'*: IWM SA 23212.

189 *His grandfather*: The details of his early life are taken from JHA, outline biography, and interview with Dick Hudson, 20 August 2009.

190 *it 'looked like'*: IWM SA 23212.

191 *a Field Park Company*: He was in 211 Field Park Company. Its BEF war diary is at NA WO 166/3680.

191 *a disorderly retreat*: Hudson's perception of it is in IWM SA 23212.

191 *'It was … astonishing'*: JHA, Dunkirk memoir.

191 *'One chap'*: Ibid.

191 *'But as the tide'*: Ibid.

192 *'We left France'*: IWM SA 23212.

192 *'You see by now'*: Ibid.

192 *After a day or two*: JHA, Army record.

192 *14 BD Company*: The company war diary for this period is at NA WO 166/4005.

193 *German pathfinder aircraft*: For the Sheffield Blitz, see MOHS, op. cit., p. 101, and *The Times*, 14 December 1940.

193 *tramcars were welded*: Mason, *Battle over Britain*, p. 477.

193 *the Marples Hotel*: Sheffield *Star*, 10 December 2007, and Gardiner, *Wartime*, p. 357.

193 *A map would have been useless*: Hudson's experiences in the aftermath of the raid come from JHA, Sheffield memoir, and IWM SA 23212.

194 *'A lot of the wardens'*: IWM SA 23212.

194 *'Experience over the whole country'*: JHA, Sheffield memoir and letter to Lord Mayor of Sheffield, 30 November 1990.

194 *'It had fallen'*: IWM SA 23212.

195 *On 19 December*: JHA, Reconnaissance reports.

195 *'We rolled the UXB'*: JHA, Sheffield memoir.

196 *'celebrated in the time-honoured fashion'*: NA WO 166/4005.

196 *'I wasn't asked to'*: IWM SA 23212.

196 *'Salmon and meat-paste sandwiches'*: JHA, letter of 26 December 1940.

197 *'I thought'*: *TV Times*, January 1979.

197 *'From it came'*: Jeacocke, *A Saturday Night Soldier*, p. 17.

197 *Three hundred people*: REMLA 9511.9.13.

198 *period of relaxation*: IWM DD 9084.

198 *'The night was bright'*: T. FitzGibbon, *With Love*, p. 64.

198 *'Curly, the Irish barman'*: Ibid., p. 65.

198 *Sunday 29 December*: MOHS, op. cit., p. 20, and Matthews, *Saint Paul's Cathedral in Wartime 1939–1945*, p. 44.

199 *'Wherever one looked'*: IWM SA 8877.

199 *'The church that means most'*: Quoted in Gardiner, op. cit., p. 361.

200 *John Donne*: Matthews, op. cit., p. 25.

200 *'I have to confess'*: Ibid., p 46. See also Gaskin, *Blitz,* p. 204 et seq. for a vivid description of the events.

200 *'And I've always remembered'*: C. FitzGibbon, *The Blitz*, p. 215.

201 *twenty-three thousand civilians*: O'Brien, *Civil Defence*, p. 677.

13: Arms and the Man

202 *'During this particular period'*: IWM SA 9400/15.

203 *'There was a great deal'*: Ibid.

203 *almost four thousand remained*: NA PREM 3/75/1, 3.1.41.

203 *'And when we explained'*: IWM SA 9400/15.

204 *'Bomb disposal'*: Ibid.

204 *'I think to begin with'*: Ibid.

204 *'The RE seemed to be heavy on Scotsmen'*: Ibid.

204 *Mervyn Peake*: Malcolm Yorke, *Mervyn Peake: My Eyes Mint Gold, A Life* (John Murray, 2001), p. 106.

205 *'When people wished to ostracise us'*: IWM SA 9400/15.

205 *'I said much more'*: Hunt, *Bombs and Booby Traps*, p. 27.

206 *'It was dangerous'*: Ibid., p. 28.

206 *'Three Nights' Blitz'*: MOHS, *Front Line 1940–1941*, p. 111.

207 *Three of these*: NA WO 166/4000.

208 *Rees, a civil engineer*: Lacey, *103 Bomb Disposal Section*.

208 *'When everyone had cleared'*: Ibid.

208 *By a fluke*: Jappy, *Danger UXB*, p. 70.

209 *It was never established*: JHA, casualty analysis.

209 *Lacey went back*: Lacey, op. cit.

209 *The following day*: NA WO 166/400.

209 *'I think it was'*: Jappy, op. cit., p. 81.

210 *'All that was left for us'*: Lacey, op. cit.

14: The Hand of War

211 *he compiled*: JHA, casualty analysis. This is a table of seven sheets dated April 1941 which sets out the details of every known UXB fatal incident to date (with one or two omissions), including the probable cause of the explosion. It also breaks these down by company and by UXB priority.

213 *lower the number of casualties*: 207 officers and men of the RE were killed in the ten months between August 1940 and May 1941, when Hudson finished his analysis. Sixty-four were killed in the eighteen months between June 1941 and the end of 1942, including some in BD work overseas. The drop in the number of UXBs, of course, also played a large part in the reduction in casualties. (The fullest record of the dead is in the memorial book now kept by 33 EOD Regiment. See, too, the appendices in Ransted, *Bomb Disposal and the British Casualties of WW2*). EODTIC UK/H8-25, Part 1 contains other statistics, including those for each BD company.

214 *Over months in storage*: Fleischer, *German Air-Dropped Weapons to 1945*, p. 73.

215 *into the path*: Museum of London, image 002724.

215 *the 'Wednesday' and the 'Saturday'*: MOHS, *Front Line 1940–1941*, p. 22.

215 *'We dug all day'*: *TV Times*, January 1979.

216 *'I was walking'*: Ibid.

217 *'And then a prostitute'*: IWM SA 5340.

217 *'Then, suddenly, it happened'*: Ibid.

217 *'This was a four-thousand-kilogram bomb'*: Ibid.

218 *It was a cloudless night*: Calder, *The Myth of the Blitz*, p. 247.

218 *the Rector*: William Pennington-Bickford.

219 *'Whitehall was thronged with people'*: Colville, *The Fringes of Power*, p. 333. Colville returned to Downing Street to be told that Rudolf Hess had landed in Scotland.

219 *'The bomb had fallen'*: Bartlett, *And Now, Tomorrow*, p. 89.

220 *'I got myself'*: IWM SA 5339.

220 *'In the main'*: Ibid.

220 *'That was the first operation'*: Hunt, *Bombs and Booby Traps*, pp. 30–1.

221 *When another landed*: Sansom, *The Blitz*, p. 77.

221 *Dozens of incendiaries*: *The Times*, 19 November 1943.

221 *Ernest Gidden*: *London Gazette*, 9 June 1942. See also *Evening News and Star*, 22 November 1960, and mentioned in Hebblethwaite, *One Step Further*, vol. 9, p. 114.

221 *He was helped*: NA WO 166/3997.

222 *being an undertaker*: Interview with Oswin Kent, 18 June 2009.

222 *In Ramsgate*: MOHS, op. cit., p. 132.

222 *It landed inside*: NA T 351/5.

223 *'I didn't fancy'*: IWM SA 14969.

224 *'We lost lads'*: Ibid.

224 *'So we got a bomb'*: Ibid.

224 *Suffolk had been told*: EODTIC UK/H8-26, Part 2. Also NA AVIA 22/1221.

225 *in precise detail*: NA AVIA 22/2254.

225 *One officer accused him*: Ibid.

225 *A few weeks later*: Ibid.

225 *'No one seemed very impressed'*: Wells, *Danger! Unexploded Bomb*, p. 42.

226 *'He was a very colourful'*: IWM SA 23212.

226 *'formed the friendliest relations'*: NA AVIA 22/1221. Godsmark had been Beckingham's officer in their trio in Sheffield at the start of the war.

226 *'while he shows'*: Ibid.

226 *On Monday 12 May*: NA AVIA 22/1221 contains the detailed report on the events at Erith Marshes. See also Ransted. 'The Death of an Earl'.

227 *'I was always a bit nervous'*: Hare-Scott, *For Gallantry*, p. 113.

227 *'a painful interview'*: NA AVIA 22/1221.

227 *The War Office had yet to pay*: NA T 164/195/22.

228 *Gough wrote an obituary*: *The Times*, 21 May 1941.

228 *'destroying unexploded bombs'*: Parliamentary Debates, House of Lords, Official Report vol. 119, no. 43, Tuesday 20 May 1941, p. 197.

228 *'a very brave and gallant gentleman'*: NA AVIA 22/1221.

228 *'My workshop staff'*: Ibid.

229 *'Some officers'*: Wells, op. cit., p. 42.

229 *'We were all very thankful'*: IWM SA 23212.

230 *not have been reckless*: See especially the witness statements of Liposta, Privett and Brownrigg in NA AVIA 22/1221.

230 *In mid-July*: *The Times*, 19 July 1941. See London *Evening Standard*, 22 July 1941 for some of the first examples of myth-making about him.

230 *'Up went the Earl of Suffolk'*: Churchill, *Their Finest Hour*, p. 263.

15: Killcat Corner

231 *'Look at today's* Times*'*: IWM SA 25426.

231 *'how very few George Medals'*: NA PREM 2/108, 22.12.40.

232 *'It was clear'*: NA T 351/1, 28.11.40.

232 *Pressure was put*: NA T 351/8.

232 *'only for acts'*: London *Gazette*, 31 January 1941.

233 *'in his view'*: NA T 351/8.

233 *'the large number'*: Ibid.

233 *General Taylor was asked*: NA T 351/8. See, too, NA WO 258/23 for general statistics on numbers of BD personnel at this stage of the war.

233 *'less consistent'*: NA T 351/1.

233 *'should be judged'*: NA T 351/8.

234 *'cold-blooded courage'*: NA T 351/9.

234 *It rejected*: Ibid.

234 *could not 'help thinking that'*: Ibid.

235 *he took his mother*: IWM SA 25426.

235 *'I happen to know'*: *The Times*, 11 June 1942.

235 *a strategic success*: Maier et al, *Germany and the Second World War*, vol. 2, p. 403.

236 *'Looking back'*: Strachey, *Post D*, p. 134. Strachey left the Communist Party of Great Britain in 1940, joined the RAF, was elected an MP, and by 1945 was Under-Secretary of State for Air.

236 *'The work was all done'*: IWM SA 23212.

237 *'I was the chap'*: Ibid.

237 *Among those watching*: EODTIC UK/H8-3 and NA AVIA 22/2263.

237 *It was not until December 1940*: For a summary of the development of the liquid fuze discharger, see NA WO 195/1911, Hurst's letters

and report of May 1945 in AVIA 22/2263 about the Stevens Stopper, and Gough, 'Research and Development Applied to Bomb Disposal'. (Hurst seems to have prepared much of what became Gough's Hawksley lecture.) Andrade's earlier work is mentioned in NA WO 195/297.

238 *'with a little salt'*: EODTIC UK/H/GE/29/1. See also Gough, op. cit., p. 12.

238 *First, the fuze*: Ibid.

238 *Cornelius Stevens*: I am grateful to Paul Hughes and Hans Houterman for biographical information about Stevens.

239 *Stevens's solution*: For Stevens's own account of the Stopper, see NA AIR 2/2902 as well as the Hurst report on it cited in the note to p. 237, above.

240 *the viscosity of brine*: EODTIC UK/H/GE/29/1.

240 *'To stop the fuze'*: IWM SA 14969.

240 *Hudson had already had trouble*: His experiences at Bristol are in JHA, Notes from defuzing a 17/50.

241 *'a narrow squeak'*: NA AVIA 22/2460.

241 *'the usual channels'*: Ibid.

242 *'it would be fatal'*: Ibid.

242 *'Fundamental doubts'*: Scott and Hughes, *The Administration of War Production*, p. 73.

242 *priority rating*: NA AVIA 22/2263, 14.8.41.

243 *'peaceful and quiet'*: Jeacocke, *A Saturday Night Soldier*, p. 15.

243 *'My batman'*: IWM SA 23212.

243 *'I remember a great feeling'*: JHA, letter to John Bartleson, 23 February 1995.

243 *'I had been given some labels'*: IWM SA 23212.

243 *'They got together'*: Ibid.

244 *'It was fascinating'*: Ibid.

16: Conduct Unbecoming

245 *'the bomb which fell'*: *The Times*, 23 February 1942.

245 *nearly thirty charges*: NA WO 84/65. The relevant Judge Advocate General's letter book and minute book are at WO 81/172 and WO 83/77 respectively. I am grateful to Peter Boalch for help in locating these.

245 *'fraud and dishonesty'*: For the details of the charges, speeches and

evidence at the trial see *The Times*, 19–21 May 1942, and *East Anglian Daily Times*, 19 May, 6 June and 8 June 1942.

247 *convicted of looting*: NA WO 166/3997.

247 *'would not be human'*: *The Times*, 21 May 1942.

247 *a statement about Jones's effects*: NA WO 84/65.

248 *As the Army lawyer … noted*: NA WO 83/77.

248 *'generosity of heart'*: *The Times*, 20 May 1942.

249 *Even the children*: Hebblethwaite, *One Step Further*, vol. 3, pp. 89–92.

249 *citation for the GC*: *London Gazette* (Issue 34956, supplement of 30 September 1940), 27 September 1940.

249 *'drove the lorry out to the Marshes'*: *Daily Mail*, 17 September 1940.

249 *while he was in the convoy*: IWM SA 2806, and Hebblethwaite, op. cit., vol. 9, p. 69. ('He sat on that bomb with his legs across it, to steady it … I saw him.')

249 *first made the assumption*: For instance, see *Daily Mail*, 16 September 1940. *The Times*'s report of the same date does not state that the bomb had a time fuze.

250 *in the month after Davies's deed*: Summary of Work Reports October–November 1940, EODTIC UK/H8-25, Part 1.

250 *late January and early March*: REMLA, UDC 623.67, Accession No. 95989.

250 *a study in the seventies*: EODTIC UK/H8-25, Part 1. The report (in 1976, subsequently revised) was by Wing Commander John MacBean, then Custodian of EODTIC. The file also contains a letter of 23 May 1942 from Brigadier Bateman to Wing Commander Lowe, which states that 'the present Ministry of Home Security instructions are based on the assumption that (17) fuzes are confined (virtually) to 500 and 250kg bombs'. Evidently, by that date it was known within BD that a thousand-kilogram bomb was most unlikely to be fuzed with a (17).

251 *'tapped it with my crowbar'*: *Daily Mail*, 17 September 1940.

252 *'The valuable psychological effect'*: NA PREM 2/108.

252 *destroyed all their records*: There is no evidence that this was not an accident rather than an attempt to cover up Davies's misappropriation of stores and money.

252 *risked his life*: The Company War Diary, NA WO 166/3997, gives several instances, for example the ticking (17) at Charringtons' offices in Lombard Street.

252 *'I told them'*: *Glasgow Herald*, 4 October 1984.

253 *one young officer*: Letter from Lt-Col J. A. Coombs to *The Sapper*, December 1984.

253 *'I don't think now'*: *Glasgow Herald*, 4 October 1984.

253 *complicated relationships*: Compare the information given in Hebblethwaite, op. cit., vol. 3, pp. 89–92 and that on his death certificate in IWM DD GC Box 4.

17: Y Fuze

255 *middle of the night*: IWM SA 23212.

255 *From the outside*: Printing & Stationery Services, *Summary of German Rheinmetall Fuzes*, p. 127.

256 *no fuze in any UXB could be trusted*: Hudson's first bulletin on the Y fuze, dated 22 January 1943, states that 'every fuze is to be suspect'. See EODTIC H/GE/20-0.

256 *'possessing outstanding ability'*: JHA, report of 23 July 1942.

256 *'Colin says'*: JHA, Gretta to John, 28 October 1942.

257 *'I wonder what'*: JHA, Gretta to John, Saturday 7.45 p.m.

257 *bombs became unstable*: Hartley, *Unexploded Bomb*, p. 130.

257 *'sudden, loud clap of thunder'*: *The Times*, 8 June 1942. See IWM LFB 103 for footage of the aftermath.

257 *'Everybody dropped everything'*: Jappy, *Danger UXB*, p. 134.

258 *'a big girder'*: Ibid.

258 *'It was just as if'*: Ibid., p. 135.

259 *The next morning*: Hartley, op. cit., p. 141.

260 *One had fallen*: The details of the Battersea UXB are taken from IWM SA 23212 and the recommendation for Martin's GC is at NA WO 373/67.

260 *The son of a Derbyshire priest*: Hebblethwaite, *One Step Further*, vol. 6, pp. 56–8.

261 *a 'dangerous job'*: JHA, draft for article.

262 *'What a marvellous story'*: NA T 351/18.

262 *'My husband's absorbing interest'*: *East Anglian Daily Times*, 12 March 1943.

262 *'We put a lead'*: IWM SA 23212.

262 *'We dripped it on'*: IWM SA 23212.

263 *'I thought that'*: Royle, 'A most disarming game of cat and mouse'.

263 *This had fallen close*: The description of the Y fuze at Albert Bridge is drawn from Hudson's own account in JHA, draft obituary. See also Hogben, *Designed to Kill*, p. 131.

265 *at a lower temperature still*: EODTIC – H/GE/29/1.

266 *'It was just after'*: Strachey, *Post D*, p. 24.

266 *'I regard any publicity'*: NA AVIA 22/2263, 11.11.40.

267 *'an excellent one of its type'*: EODTIC H/GE/24.

267 *'liquid air was being used'*: Ibid.

267 *'learnt that these'*: EODTIC UK/H/8-3. This is the UXBC's summary of work done in 1943.

267 *'John my very own beloved'*: JHA, Gretta to John, 26 January 1943.

268 *'I was so excited'*: JHA, Gretta to John, Good Friday 1943

268 *'Colin didn't cry'*: JHA, Gretta to John, Monday 8 o'clock

268 *'It's impossible'*: JHA, Gretta to John, Saturday 7.40 p.m.

18: Butterflies

270 *'What I liked'*: NA SA 14131. The information about Shoebottom comes from the same recording.

270 *the butterfly bomb*: Hogben, *Designed to Kill*, p. 117, and Hunt, *Bombs and Booby Traps*, p. 37.

271 *Ray Bingham*: Hunt, op. cit., p. 45. There are several versions of the circumstances of Bingham's death. Hunt's has the most detail.

272 *obsession with security*: NA WO 32/9760.

272 *In the early hours of Whit Monday*: The details of the raid are taken from the *Grimsby Telegraph*, 24 June 2003, and from the official reports of 19 June 1943, July 1943 and 16 August 1943 in EODTIC UK/H8-24.

275 *'A bomb complete with wings'*: IWM SA 14840.

275 *'I put a bit of string'*: NA SA 26831.

275 *'The worst one'*: IWM SA 14840.

276 *'Twelve of them'*: *TV Times*, January 1979.

276 *'It took three months'*: Ibid.

277 *'now ... we have to fear'*: *Fall of Anti-personnel bombs in Grimsby*, 19 June 1943, EODTIC UK/H8-24.

277 *'In view of our'*: *Report on the raid on Grimsby*, July 1943, EODTIC UK/H8-24.

19: Hitler's Last Hope

275 *Forty-two thousand people*: Lowe, *Inferno*, p. 206.

278 *'a weapon with which'*: Quoted in Gardiner, *Wartime*, p. 638.

279 *The rocket*: See Hogben, *Designed to Kill*, p. 158, for more on the V-1's specifications.

280 *'In the distance'*: Sheppard-Jones, *I Walk on Wheels*, p. 12.

280 *'I knew its warhead'*: Longmate, *The Doodlebugs*, p. 122.

281 *'A particularly nasty'*: Demarne, *The London Blitz*, p. 119.

281 *The charge sat forward*: Hartley, *Unexploded Bomb*, p. 186.

282 *'every possible step'*: JHA, *UX Flying Bomb* report by Hudson himself is the main source for this episode. See also NA WO 373/68 and *London Gazette*, 15 September 1944.

282 *'It must have occurred'*: IWM SA 23212.

238 *'Bob and I decided'*: Ibid.

283 *'We went to a girls' school'*: Ibid.

286 *'I feel very sincerely'*: Rose, *Elusive Rothschild*, p. 71.

287 *'Oh John'*: JHA, Gretta to John, 29 July.

20: Sing as We Go

288 *'The procedure'*: *TV Times*, January 1979.

289 *'I straightened'*: Ibid.

290 *'I think I was run down'*: IWM SA 23212.

290 *Clearing the country's coast*: Hartley, *Unexploded Bomb*, op. cit., p. 236.

291 *About 147,000 British civilians*: O'Brien, *Civil Defence*, p. 678.

291 *68,500 tons of explosive*: EODTIC UK/H8-24, Ministry of Works letter of 28 November 1949.

291 *a standing start*: There are several measures of the numbers of UXBs defuzed. I have used the figures in the reports prepared for the government both during and after the war (the latter by Herbert Hunt) contained in EODTIC UK/H8-24.

291 *a conservative estimate*: See the note to p. 213, above. I have included in this figure only those killed clearing mines before the end of 1945, but several dozen more were to lose their lives in the same fashion by 1947. The UXB fatalities suffered by the Royal Engineers between 1940 and 1945, as recorded in 33 Regiment's memorial book, totals 603. This includes those who died abroad while on BD duty. Ransted's *Bomb Disposal and the British Casualties*

of WW2 lists the names of some 120 RN and RAF personnel killed disposing of UXBs.

292 *'in accord with the ideals'*: NA AVIA 22/2263. The copy of the interrogation of 'Julius Renner' at EODTIC H/GE/57 is probably that of the same soldier.

292 *At first there were examples*: For details on the Germans' use of camp prisoners see Blank et al, *Germany and the Second World War*, vol. IX/1, p. 249 et seq.

293 *'10 per cent'*: EODTIC H/GE/57.

293 *As in Britain*: Hartley, op. cit., p. 226.

294 *Documents found there*: Merriman's papers are at IWM DA 69885.

294 *a new file*: NA AVIA 22/2263, 29.9.45.

294 *Bob Hurst*: Obituary, *The Times*, 7 June 1996.

295 *'When you're coming back'*: IWM SA 25426.

296 *'My wife told me'*: IWM SA 23212.

296 *He may perhaps have regretted*: Interview with Dick Hudson, 20 August 2009.

296 *'an engineer of the highest calibre'*: NA SUPP 20/5, 31.5.45.

296 *'one who is likely'*: Ibid.

297 *'I hope that you will be able'*: NA SUPP 20/5, 17.8.45.

297 *The most celebrated*: Edgerton, *Warfare State*, p. 163.

297 *'made available to us'*: NA SUPP 20/5, 16.8.45.

297 *'The Director General'*: Scott and Hughes, *The Administration of War Production*, p. 286.

298 *'enthusiasm and undaunted cheerfulness'*: NA SUPP 20/5, 16.8.45.

298 *a round of golf*: AIME, COP 18/6, letter from J. R. Gough, 3 June 1965.

299 *'A typical case'*: Rühlemann, *Father Tells Daughter*, p. 390.

299 *'really had nothing'*: Ibid.

300 *He did not forsake*: Interview with Elga La Pine, 25 March 2010.

302 *'I wanted to try'*: IWM SA 20693.

302 *'just an ordinary person'*: Obituary, *Daily Telegraph*, 26 April 2002.

302 *'I dithered'*: *TV Times*, op. cit.

302 *'I didn't know'*: NA SA 14131.

302 *'You was always'*: Ibid.

303 *more than two thousand tons*: *Der Spiegel*, 14 October 2008.

303 *At Christmas 1958*: Hogben, *Designed to Kill*, op. cit., p. 207.

304 *a map to be distributed*: *Daily Telegraph*, 13 July 2008.

304 *During construction work*: *Daily Telegraph*, 5 June 2008.

304 *'the bravest of the brave'*: *The Times*, 19 March 2010.

Afterword: The Whole Earth

306 *'In my view'*: IWM SA 25426.

306 *And while committing*: Thucydides, *History of the Peloponnesian War*, Book II, in the translation by Richard Crawley.

INDEX

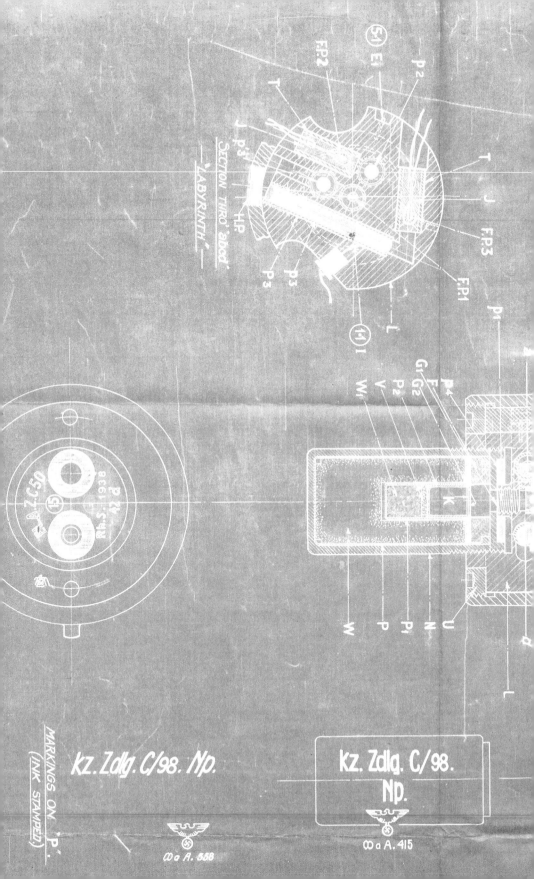

SECTION THRO' 'abcd'
"LABYRINTH"

P2
FP2
E1
5:1
P2
T
J P3
H.P.
P3
P3
J
T
F.P.3
F.P.1
L
P4
1:1
I
G1 G2
F
P2
V
W1
P1

K
W
P
P1
N
U
L

E A 7.C.50
5:1
Rh.S. 1938
d 43

kz. Zdlg. C/98. Np.

kz. Zdlg. C/98.
Np.

Wa A. 558

Wa A. 415